Me and THE BIOSPHERES

A Memoir by the Inventor of Biosphere 2

JOHN ALLEN

SYNERGETIC PRESS Santa Fe, New Mexico

Published by Synergetic Press
1 Bluebird Court, Santa Fe, NM 87508

Library of Congress Cataloging-in-Publication Data

Allen, John (John Polk)
  Me and the biospheres : a memoir by the inventor of Biosphere 2 / John Allen.
      p. cm.
  Includes bibliographical references.
  ISBN 978-0-907791-37-9 (alk. paper)
  1. Biosphere. 2. Biosphere 2 (Project) 3. Allen, John (John Polk)
  I. Title.
  QH343.4.A55 2009
  577.092--dc22
  [B]
                                              2008000943

Cover photograph of Biosphere 2 lung construction by Goya R. Kenny
Book design by Arlyn Eve Nathan
Production editor: Linda Sperling
Photo editor: Deborah Parrish Snyder
Typesetting by Deborah Parrish Snyder and Monique Sofo
Typeface: Janson
Printed by Arizona Lithographers, Tucson, Arizona USA

To my Grandfather and Anil Thakkar, the two greatest men I ever met, one from the West and one from the East who showed me how to be a biospherian. And to everyone who helped and to some of those who hindered.

Table of Contents

Sustainable forestry project,
Puerto Rico, 2008

Preface

Before I was born out of my mother generations guided me,
My embryo has never been torpid, nothing could overlay it.

For it the nebula cohered to an orb,
The long slow strata piled to rest it on,
Vast vegetables gave it sustenance,
Monstrous sauroids transported it in their mouths and deposited it with care.
All forces have been steadily employ'd to complete and delight me,
Now on this spot I stand with my robust soul.

WALT WHITMAN

I DELIGHTED IN THE REALITY of what I later learned to call Nature. I smelled the fresh-spaded dirt in spring and ran barefoot over grass and clay and small rocks. I slid into sun-warmed summer wallows with the big tough-skinned boars and sows that snuffled about in my grandfather's spacious hog lot. I loved the big-branched tall cottonwoods that flanked the quick-sanded river, from whose safe heights squirrels cautiously peered down at me. I sat contentedly watching the stinging red ants scurrying in and out of their sun-baked, barren mounds that hid the queen.

I dazzled myself by chasing butterflies and running down dirt roads, chased by a cloud of midges. I dipped fruit jars into ponds to catch polliwogs and watched them metamorphose into frogs, picked potato bugs off my grandmother's green plants and marveled at their increase, seined for crawdads in small creeks. I sat on my haunches for hours with Grandmother's Rhode Island Red chickens, listening to the hens bragging when they laid an egg and watching the glossy-feathered roosters strut around, their proud combs waving.

I felt connected to every living thing eating and being eaten, and breathing in and out the same air. Worms fascinated me because they would eat me after my death. Sharks and tigers fascinated me because they would coolly eat me alive, if they could catch me, enjoying the same hunger-satisfying delicious taste I did when I ate the crisp thighs and tender white meat of a chicken. My grandmother could wring the necks of two chickens at once, whirling one in each hand in opposite directions. Those chickens ran around the backyard with frantic energy, spouting blood until they fell over. She boiled the chickens in hot water in a big iron pot until they could be easily plucked, and cut the brains out of the heads to serve with scrambled eggs.

The men around me bored me when they talked too long about cars – Fords versus Chevrolets –but I tuned in to every knowledgeable word when they talked about crops: when to plow, harrow, harvest, what to plant, the differences between corn and wheat and rice and barley and millet and sorghum. Grandfather told me always to tithe ten percent of any crop I grew to the insects because other beings wanted to eat, too. The men talked about which animals to have around and the differences in cattle – Shorthorns, Whiteface, Black Angus, Holsteins, Guernseys, and Jerseys – and why Grandfather chose a Jersey over all the other cows to milk every day but a Whiteface for beef, while my uncle Frank Ball chose Holsteins for his dairy farm to ship milk to Oklahoma City. I heard about how to judge breeds of horses – Thoroughbreds, Quarterhorses, Morgans, Tennessee Walkers, Appaloosas, and reservation Indian horses. They taught me to tell a rattler from a bull snake and a racer from a garden snake, to appreciate the difference in habits of a chipmunk and a squirrel, and in temperament between bass, trout, catfish, perch, and minnows.

My father was the scion of two famously idealistic lines, the Allens and the Cutwrights; my mother came from the Polks and the Walls, both renowned for their no-nonsense style. The Allens came via covered wagon into the Indian Territory from Kansas, always staying north of the Mason-Dixon line on their way West; the Walls arrived by horse and covered wagon, always staying south of the Mason-

My English-Scotch-Irish
parents, Paul and Opal Allen

John, age twelve (sixth from left), father, Paul, and younger brother, Roy, (far left) with extended family at Uncle Frank and Aunt Sarah Nee's farm, Paul's Valley, Oklahoma, 1941.

Dixon line. My uncles and aunts liked to point out the differences in my two family lines: the Allens, tall mesomorphs, liked idea systems; the Walls, shorter but with even broader shoulders than the Allens, were earth- and people-oriented and liked to puncture pompous balloons. Allens and Walls both shared frontier values of hardiness, thrift, growing your own, individual independence from an early age, and moving on to new frontiers when the old one had been moved in on. The culture that my forefathers created proudly called itself Oklahoman.

To the north of us, beyond two rivers (the North Canadian and the Cimarron), lay the border that separated us from the culture called Kansas. South of my Grandfather's wheat and cattle farms along the South Canadian River, beyond two more rivers, the Washita and the Red, lay the border of a culture called Texas. These two sub-cultures, Kansas and Texas, had spun off from the North and the South when these two sections engaged in a great race to beat each other to the Pacific Ocean in order to control the American State.

To me, my uncle, Bus Wall, was an acknowledged authority since he had been a .400 hitter and a straight-A student at the University of

Oklahoma. It was he who told me that the last person who could hold these two cultures together without fighting had been James K. Polk, my great-great-great-great uncle. Polk had brought America to the shores of the Pacific from San Diego to Seattle and brought Texas into the Union (although Texas had betrayed the Union despite the family's hero, Sam Houston, opposing Texas's joining the Confederacy). After Polk, said Uncle Bus, nobody could hold the Divided House together, and soon afterwards the two cultures fought the Civil War.

Uncle Bus worked as a firefighter in Oklahoma City rather than, in his words, as a "bought and paid-for intellectual." Uncle Bus said Oklahoma culminated the American dream because Oklahoma culture synthesized the North and the South, via Kansas and Texas, as well as the great plains of the West. Our Great Plains climate and geology fostered a vaster view of things, he continued, since trees grew mainly in the river bottoms, and no mountain ranges blocked the view to turn people's vision up to pie-in-the-sky.

In addition, Oklahoma meant "home of the Red Man." More Indian tribes lived in Oklahoma than in any other state. In fact, the greatest Oklahoman had been a Cherokee, Will Rogers. There was a bit of Indian in most frontier families because the men had gotten out ahead of the women and married Indian women, as had my great-grandfather Buck Wall, so Uncle Bus told me. Uncle Bus proudly declared that Oklahomans had written the longest state constitution specifically to limit state power, exemplifying all the best principles of the two greatest figures in world political history, Thomas Jefferson and Andrew Jackson. Jefferson wrote the theory and constitution of a sustainable democracy; Jackson put it into practice. Under the powers of its state constitution, Oklahomans had impeached three rotten governors in the first thirty years of its existence, whereas other states had to suffer scoundrels till the next election.

Above all, I never grew tired of hearing of why Western Oklahoma was different from Eastern Oklahoma (lower rainfall), why South-eastern Oklahoma was called Little Dixie, why the Panhandle was different because it was the cattle-dominated dry high plains out past the short grass wheat country. Though I didn't know the word then, all that windswept, flash-flooded talk felt like sheer magic – it always lassoed and calf-roped its target. Later I learned that kind of talk shared with pure poetry direct perceptions of Nature – a sunburnt language that integrated fact and symbol.

I was intrigued by everything that lived underground in the soil: the night crawlers' trails of dried slime gleaming in early morning sunshine, twisted roots, mushroom threads, tireless blind moles, and farseeing prairie dogs sitting alertly at the entrance to their escape holes. I strolled by the riverbank and gazed at the strong bare roots of great cottonwoods which still held the trees upright although stripped of the protection of several feet of earth by the recurrent, remorseless floods brought on by the use of the mold board plow and the destruction of the deep intertwined grass roots.

At the age of nine, I learned how to use the public library. For good or ill, I devoured authors who taught me that the world was divided into art, science, philosophy, religion, physics, chemistry, biology, culturology, linguistics, astronomy and history, and on and on into books, chapters, paragraphs, sentences, and phrases. I decided to master them all – at least to a level where I could hold a conversation with the past masters of each specialty – and to develop three areas where I would be a master and maybe write one of these magic books.

And I never tired of barefoot, bare-chested walks in winter green wheat grass or hot yellow August stubble or loafing by those flood-cut banks. I really loved being natural and hated the shoes that tightened my toes and kept me from touching Mother Earth. Whenever I could get away with it, even in winter, I would wear only a T-shirt or strip it off on milder afternoons to enjoy the bracing cold. The rabbits and the squirrels lived in the cold, why couldn't I? Walking through sweet

John at age ten started a two page weekly newspaper, the Chit-Chat, *reporting news of world events. It had fifty subscribers and sold for two cents a copy.*

flowering purple fields of alfalfa, I would reach the tangled bank of the river and sitting on some flood-delivered log, watch water spiders skim on quiet pools, crows flap, and keep an eye out for the rare Baltimore Oriole with its telltale flash of orange. Later I learned that all those objects of watching could be treated as subjects for biological science, but I wasn't really excited by old Mrs. Dolan when she had us dissect frogs and memorize the names for parts of dead animals.

Age eleven in Okemah, Oklahoma.

When I turned eleven, Mars loomed large and red in the sky; reading the Martian stories of Edgar Rice Burroughs, I dreamed of beautiful Deja Thoris. I would walk to the edge of Grandfather's open porch, stretch out my arms to Mars, and wait faithfully to be transported to that planet, like John Carter had been. When night fell, I would throw myself on my back to gaze at the Milky Way, watch the bats flit by, and hope to join Sitting Bull, Crazy Horse, and Quanah Parker up there in the Happy Hunting Grounds. Three of the original astronaut corps told me they had also tried to do the same thing as John Carter.

I always considered *me* – the part of myself that was so connected to Nature – as much smarter than *John*, who was always reading textbooks, getting to school on time, dressing warmly, making alternate All-State in football and A's in every subject, and getting lost in crazy make-believe dreams about delightful girls. *Me* found it more interesting to listen to the fluttering leaves of the cottonwoods and the agile squirrels. *Me* always took the side of Huckleberry Finn, not Tom Sawyer. *Me* preferred Bomba the Jungle Boy to Tom Swift and his Electric Car. However, *me* would let *John* read whatever I wished. John was a very interesting butterfly or hummingbird to *me*.

What *me* really liked was the whole living, dying, changing, highly differentiated, intricately connected world – the great biosphere of Earth. And nothing much has changed in all the years since I ripened into Age.

The Music of the Spheres

CHAPTER ONE

At the same time collecting and recollecting Memory
For that Shipwrecked moment
When we need it all and then some
To keep the voyage going going
And never gone.

JOHNNY DOLPHIN

ONE DAY I LOOKED OUT of my Manhattan skyscraper office at Number One Whitehall Street toward the Statue of Liberty and realized I couldn't open my window. Some liberty. Multiple levels of meaning packed into that realization shivered my timbers. I stood bolt upright and thought:

I don't have the power to open any window right now in my own skyscraper of a body. I can't catch onto anything that's not branded and packaged. If the scholarship's not Harvard, the theater Off-Broadway, the girl Vassar, Sarah Lawrence, Hamish Bohemian or Black Rebel, my buddy not Corporate Entrepreneur or Underground Poet, my own thinking post-Aristotelian and post-Marxist Deconstructionist, if the science doesn't have a chair at Oxford or an invention-crammed garage or the book's not classic or avant-garde, I won't touch it. Those are all gorgeous windows in my shining skyscraper giving me million-dollar views of the world's splendid traffic, but I can't open them to smell the fresh air or hike barefoot along an open road or, worst of all, walk away from a pompous power holder. Get me outta here!

At that exact moment I saw a Yugoslav freighter sailing between Brooklyn and Staten Island, heading out across the Atlantic for Tangiers in North Africa. Although I lived in a great pad in Little Italy south of Houston Street (before it became Soho), hung with a lively cultural crowd at the White Horse Tavern and the Kiwi Bar where you never knew who you'd meet, love, or fight with, and took wild well-paid expeditions to Iran and the Rocky Mountains and

John on board the Hrvatska.

IMAGE ON LEFT: *Passengers on board the Yugoslav freighter,* Hrvatska, *headed for Tangiers in 1963.*

Age fifteen, Venice Beach, 1944. "I worked at Helms Bakery for sixty cents an hour, six days a week."

Plateau West, I knew what I had to do next–get on the next Yugoslav freighter headed East, to Tangiers. I had people to see and places to go where I had not the slightest idea what would happen next.

It was not the first time wanderlust had seized hold of my life. When I turned fifteen, I wandered the West Coast from the Mexican border cities to the Yakima and Wenatchee valleys of the Northwest, working in cotton, oranges, potatoes, apples, and timber. Some may have called me a lowly migrant worker, an "Okie" refugee from the "'dust bowl" caused by that huge over-plowing, but this outdoors existence filled me with revelations. A tireless glory of deserts, irrigated valleys, grasslands, forests, cliffs, timberlines, and salmon rivers showed me the richness and productivity of life and the boundless ingenuity of humans.

My intellectual wanderlust outdid the physical. On reading at age four-teen Alexis Carrel's *Man, the Unknown*, I exploded with never-ending energy. Carrel maintained that if a person resolutely started to master one of the great divisions of knowledge every two years from the age of fourteen until forty, at that point one could begin to really live. I followed his advice until I turned thirty-four, going through box after box of knowledge with religious discipline, with great vigor and success. Going from box to box was a rational malarkey with a double bind – boxing myself in and boxing myself out. I had learned very early in Granddad's hog wallows how to live – relax in the summer sun and green-scummed pond with a grand old boar, and ease into whatever scene comes up, with plenty of moxie. However, with increasing skill in manipulating the upper lobes of my brain and increasing neglect of the ancient limbic and oblongata wisdoms of direct emotion and sense wisdom, I heard myself laugh less and less often.

By the time I decisively left, boarding the freighter on the Brooklyn Dock for Tangiers in October 1963, I had attended five universities, graduated from two, read innumerable books, been consistently rated super student, awarded scholarships and honors. I held instructive jobs – picking fruit in apple and apricot orchards, lumberjacking, working as a skilled machinist, in assembly line mass production, union organizing in Chicago, as a machinist in the Army Corps of Engineers, running a farm, setting up corporations, discovering a huge deposit of coal, and whizzing around on helichoppers and in Hertzes doing ambitious technological-political projects. I enjoyed tumultuous relationships with bright-eyed women, made pals with adventurous men, published a few poems, got rated as a success story by my friends. I was moving steadily forward in my three selected fields – the "boxes" or "monads" of enterprise, theater, and biospheric geology. I had gone bonkers with multi-phrenic intensity.

I hoped to learn to prolong and deepen my first two tastes of unity in diversity. In the summers of 1962 and 1963, two shamanic experiences changed me forever. A friend of mine, with whom I met some real Indian leaders in the Southwest, had become an initiate in peyote mysteries and rites. I joined him in his next exploration. As a B-52 winged overhead, I "saw" that the pilot, disciplined like myself, would just drop his atomic bombs whenever and wherever ordered. I "saw" transience, death, and realities of being that made me love and surrender to the worlds of life. I decided to leave the magnificent shambles of our new Babylon.

So I turned back to *me*, and to the biosphere – the real love of my life. I sailed on the Yugoslavian freighter *Hrvatska* for Tangiers, which had become a center for the avant-garde's refreshing encounter with Berber magic and music in its fabled "Interzone." I wanted out of all the boxes, out of all the fenced fields; I wanted to start roaming great nature, the cosmos.

I headed for Morocco to start the first leg of my journey to integrate my being with me, myself and I, all once again in the right order, as my grandfather – who embodied the quintessence of frontier wisdom

Terry Taylor and John at the Socco Chico Café, Tangiers, taking in the scene.

– and my great cowboy cousin, Buck, would have put it. I eventually wrote a novel, *Thirty-Nine Blows on a Gone Trumpet* (taking about twenty years for revisions), about a character named Madison, who escapes his American fate by going to Tangiers, then a second novel about Madison's further adventures called *Journey around an Extraordinary Planet.*

Further experiences with individuals whom I can best describe as wizards of intuition who hung out in Socco Chico (that magic little plaza in Tangiers), made it clearer to me that once anyone thought of reality as existing in separate boxes, a stealthy hypertext of fishy stories stole all one's attention. This mobile palimpsest – created with such snakelike cunning, cleverly squirreled away in such separated holes, howled about by such crafty coyotes, bellowed about by such big bulls, cackled about by such proud hens, bleated about by such sad sheep, and butted around by such belligerent goats – made me jump quicker and quicker from box to box in a vain attempt to corral reality. But I had never jumped out of box-land until I met the Native

American mysteries, followed by meeting the Berbers. Think outside the box from within another box or from between boxes? Not the chance of a snowflake in hell.

I had begun to catch on about how to relate *me* to the biosphere in 1953 when I took a course in Historical Geology taught by Professor Ben Parker at the Colorado School of Mines. Scorning all rhetorical devices, he stated flatly that around the great spherical core of the Planet Earth there existed a sphere of rock, a sphere of air, a sphere of water, and a sphere of life – a biosphere. He further asserted that all these spheres' interrelated history produced constant and vast changes. I felt the famed Pythagorean "music of the spheres" thrill through me, and for the first time felt that I might amount to something, that I had found the clue to a creative life. This biosphere idea dealt effectively with sense impressions and emotional assessments, fit with what I intuitively grasped while walking by myself up Clear Creek Canyon near Golden, Colorado. All of life – undissected, radiantly alive, extending over the whole earth in its waters, soils, and airs – formed a single, whole evolution, connected throughout time, space, matter, energy and quanta with the evolution of the entire cosmos. All my culture's money-laden, double-bind boxes momentarily dissolved in the belly-throbbing laughter of this sublime, diversely manifesting, evolving intelligence.

Hearing Professor Parker enunciate the co-evolution of biosphere, atmosphere, hydrosphere, and lithosphere (life, air, water, and rocks) immediately cleared up a hundred apparently disconnected problems. Problems of location throughout space and time of coal deposits, iron mines, precious metals, and other outstanding features of the landscape, called biomes – such as forests, grasslands, deserts, marshes, and coral reefs – fell into place. Instead of being artificially separated in boxes by political boundaries, textbooks, and corporations, these ever-changing entities formed a single four-dimensional map with moving continents, mountain ranges, and oceans, always changing position relative to the sun and galaxy in the process of organizing trillions of tons of intricately looped molecular flows that continually transformed a planetary-scale volume of space.

Me had been right all along, but only now could I speak about it, communicate and learn with other people. I began to write new poetry to purify the semantics of what *me* knew, did new science that gave *me* a kick, wandered anywhere and always found myself at home because anywhere was part of the biosphere. Home is where the heart is!

But the first thing I had done with the knowledge that the biosphere existed as a gigantic, ancient, and ever-changing entity was to unite that objective knowledge with moment-by-moment being-perceptions made by *me*. I started hanging out with rock formations deposited by long-gone biospheres as well as with the plants and animals doing their geologic duties today. I became something of a whiz at using traces of past biospheric periods to search out ore deposits and discovered a whole new way to explore for minerals, which culminated in my 1962–63 coal explorations for David Lilienthal's Development and Resources Corporation.

David Lilienthal in the early 1960s, my mentor on regional development; I worked on projects for him in Iran, the Ivory Coast and Haiti.

David, the former first chairman of the Atomic Energy Commission (and before that the Tennessee Valley Authority), became my greatest teacher in starting up and managing complex projects. In his memoirs, David called me "a crackerjack engineer." Eventually, I found four hundred million tons of unclaimed high-grade coal. David only wanted two hundred million tons to secure his standing in the energy industry when he made his major speech questioning the rise of atomic power. When I asked, he kindly permitted me to lease the other two hundred million tons, and with two Japanese investors, I became a paper billionaire (royalties would be ten dollars a ton).

Soon I grew to understand the implications of this unbridled fossil fuel exploitation. The place, Kaiparowitz Plateau, where I had found the coal, was fragile, high, cold desert ecology sacred to Hopi and Navajo cultures. Selling Carboniferous Era deposits would lead to endless trainloads stoking huge furnaces, generating ephemeral money, trailer towns, electricity for frivolous uses, and sulfur and carbon dioxides pumping smog into the pure air of the Four Corners high plateau (the meeting point where four states, Utah, Colorado, Arizona and New Mexico, come together). I dreamt of becoming rich, powerful, and artistic, then at fifty or so, doing again what came naturally. When I realized I had it all backwards – selling my youth and honor for a chancy old age – I gave up my extensive leases and terminated my company, Mountain and Manhattan. I needed to journey around and experience this extraordinary planet to the fullest. I didn't want to shuffle jejune formulas of greedsters who thought they knew what made the world go 'round. I needed to educate myself: *Educe*, lead myself out – rather than *Se-duce*, lead myself astray.

Earth *is* an extraordinary planet. A biosphere made its trillion-tonned, fifty-million specied, four-billion-year-old home here. It deposited a quadrillion tons of synthesized molecules to improve that home, and it adds ten million more tons of new geological deposits each year, thus greatly altering Earth's crust by creating rock-destroying roots and hooves, soils, mucks, sedimentary formations like coal and limestones, and liquid oils and natural gases, finally by evolving humans who evolved shore-changing and mountain-leveling technologies. Ordinarily, planets don't have biospheres. This biosphere, unique in our solar system and for much farther beyond, allowed and assisted us to evolve and to create what I now call the *ethnosphere*, the sphere of cultures that manifests differing aspects of human potential. The biosphere not only grew and evolved, but humanity evolved and developed these varying interacting cultures.

All these cultures – past, present and those to come – have given us humans awesome abilities to adapt to rapid changes. These mysterious powers ensure humanity can deal with any physical test that hits us, short of a cosmic catastrophe. For example, a cosmic catastrophe could come along due to the ethnosphere's own success: the degradation of Earth's biosphere by several of the panoply of cultures that overemphasize, even worship, the technosphere and which will not so far give up their potential to make a great atomic war or to industriously devastate the life systems.

Just as Ben Parker awakened me to the biosphere in 1953, so Melville Herskovits, at Northwestern University, had awakened me in 1946 to the ethnosphere. In his anthropology class, Herskovits baldly stated the "Boas and Benedict proposition" – that a vast arc of human potentiality existed, that each culture represented at most only a few degrees of that arc. I decided to make every effort to observe and experience in my own particular way every degree in the total arc of that sphere. Eventually, I saw that this arc, begun over forty thousand years ago, had completed its journey around the planet and twined tendrils around each human body-mind. My paper on the ethnosphere which I wrote in 2002,[1] presents my thinking on this great offspring of the Earth's biosphere.

After six years of marvelous adventures that started immediately upon boarding the *Hrvatska* for Tangiers in 1963, I found myself ready, able, and willing to live at the pitch of comprehensive action. I was no longer satisfied to poodle along, rapturously tramping about in lunatic mystic unity with deserts, jungles, and far-out cultural achievements. I wished to move my center of being from neurosoma to neurobrain. I now wished to compose a reality script based on direct perceptions emerging from centering on the actual flows in my synapses and give them the chance to receive inputs from all forms of reality. I trusted that flowing with these flows would, with the now subconsciously embedded structure of my previous knowledge about

John in a moment of contemplation, Vietnam, 1965.

role-playing and staying healthy, allow me a deeper synergy of my three lines of action: theater, enterprise, and ecotechnics. I hoped to align my life's activities with biospheric evolution.

First, I determined to help develop a complete science and engineering discipline (which I called *ecotechnics*, the ecology of technics and technics of ecology) on a biospheric-geospheric-solarspheric scale. To accomplish this, I needed to work out a conceptual model of the biosphere and then actually make a large-scale experimental model which could include humans. Second, I wished to help create the techniques and base for a theater, exploring the possibilities inherent in humanity indicated by looking at all cultural manifestations of biosphere – past, present, and future. Third, I wanted to make beautiful and sustainable project-enterprises around this lovely planet that harmonized ecology with economics. I hoped that a self-organizing way of life attuned to the biospheric rhythms would emerge for me, my friends, other humans, plants, and animals by judiciously interweaving biospheric, technical, and cultural processes in practical projects I called "synergias."

I saw the possibility of a creative alternative to destructive development, namely, *comprehensive co-evolutionary anticipatory design*, I endlessly practiced making trial integrations of the four great vectors of rocks, air, water, and life–and of their children, the soils and mucks –into a unified perception that could bring forth a model experiment whose results would ground a real way of life. My exercises expanded to include so much time, space, material, and biospheric complexity that I began to "see," to "feel," and then to constate my own eating, elimination, and physical movement as an active part of everything acting within the biosphere. Perceptions that integrated the vectors of earth, life, cultures, and technics could enhance any project. This vision of a science of the sciences gave new breadth and depth to my loves and friendships, to my poetry and stories, to my contemplations of human destiny.

I developed ways of using the hypothesis-free idiographic method of naturalist observation to complement my detailed hypothesis-driven studies relating to biospheric phenomena, demanded geologically by the mining and metallurgical departments of Colorado School of Mines and by my professional life. Konrad Lorenz, the founder of ethology, told me that only this experience-based method could generate a large enough empirical base to keep wishful fantasy out of gestalt intuitions. In addition, without continuing naturalist or discovery observations that increase the base of data, hypotheses working and re-working their givens dwindle into "the soup of the soup of the soup of the duck (whatever system they are studying)."

Also, without the feeling of personal connection with the cosmos, the work and consequences of hypothetical scientific effort disappear into a lumpen intellectual professionalism. Concealing and/or ignoring "unpleasant" data become rife. Curves get "smoothed out," vital data misplaced in miscellanies of detail. Applause replaces appraisal, criticism replaces critique in the scientific journals. Back scratching replaces head scratching.

Rangy Ed (Edward O.) Wilson added the knowhow in his enlightening book *Biophilia* that these idiographic operations could best be carried out in a state of consciousness he called the "naturalist trance," in which one can "see" key connections producing the observed phenomena. In Vienna, Lorenz, sitting by my side in front of his aquarium, showed me exactly how to fine-tune these operations when he said:

"Find your own way to attain the naturalist trance, stay in it until you see something new, then record that observation. Classify and meditate on the data until patterns emerge."

Acute Jane Goodall told me that she had also learned this observational skill from Lorenz, which she applied so brilliantly to her ground breaking studies of chimpanzees.

I wanted to observe the biosphere the way Lorenz observed geese, Wilson observed ants, and Goodall observed chimpanzees. More, I wanted to imprint the biosphere the way the goslings imprinted Lorenz, as my mother. Biosphere 2 began to exist in my mind – inchoate, inescapable, and ineluctable – beckoning me towards destiny with mysterious but never-failing allure.

Synergia Ranch, Santa Fe

CHAPTER TWO

The biosphere is a geological force and a cosmic phenomenon.
VLADIMIR VERNADSKY

Synergy is the property of a total system that cannot be predicted by addition of the properties of its parts.
R. BUCKMINSTER FULLER

FROM THAT MOMENT IN 1953 when Ben Parker's calm, nasal voice of truth stunned me into the trance of wonder, I wandered for sixteen years throughout many of the biosphere's varied bioregions–ten years at some of the living tips of technological prowess, followed by six years of on-the-edge expeditions to search out artistic and physio-psychological intuitions and practices kept alive by Earth's ancient cultures. Those travels deeply rooted and fertilized the notion of biosphere in me; they also got me a partner in biospherics.

Marie Harding and I met in India when our two journeys around the biosphere first crossed in 1964. We made our first small expedition to the Qutub Minar and Asoka's wrought iron pillar in old Delhi under the guidance of the redoubtable sage, Anil Thakkar. Marie was staying with a fellow classmate, Wendy Green, daughter of the number two man in the American Embassy, Marshal Green, a diplomat famous for his understanding of Asian cultures. Wendy had been a fun-loving Sarah Lawrence date of mine in the old Manhattan days.

A few months later (after traveling in Nepal, Tibet, Burma and Thailand), I ran across Marie in Saigon at a library, where she was relaxing on a short break from her work at a one-doctor, three-nurse hospital located in the jungle three hundred miles to the north along the Ho

IMAGE ON LEFT:
The Ortiz Mountains and Synergia Ranch, 1975.

John, back from Special Forces patrol as a Stringer Correspondent, and Marie at Project Concern, Vietnam.

Chi Minh Trail. Marie had been choppered down to Saigon and would be choppered back to the valley of the River of Dreams by a couple of spooks who liked to practice their parachute jumping where the hospital was located in the center of the rugged Montagnard tribal country. This was a mountainous, forested region where the tribes still used crossbows, and whose lands both the Commissars of the North and the Capitalists of the South coveted.

She'd said, "Come up and see me," like it was an invitation to get on the subway and visit her on the Upper East Side of Manhattan. Her family united old-line New England textiles and two centuries of Harvard with New York Finance and Upper East Side society. Marie had, in fact, graduated from Miss Porter's School, attended Sarah Lawrence receiving her BA as an artist, and then studied at the New York Art Student's League. Several of the top teachers had recognized her talent; she could catch the essence of the most varied characters in a revealing pose or scene, integrate their timelines into an all-revealing eternal moment and fix it like a butterfly on a gessoed canvas. As a kid, she was a tomboy and earned the moniker of Marie the Marauder. She had worked with a Swedish health outfit in South India before traveling on to Montagnard Vietnam.

The doc at the hospital, Jim Turpin, was a great-hearted and charismatic man who could use technical help, Marie said. I had gotten my MACV (journalist) permit as a stringer for the *Far Eastern Economic Review* by agreeing to keep my beard trimmed and tennis shoes clean, so, as a correspondent, I was free to move throughout South Vietnam to all U.S. military facilities, although I preferred life at the Majestic Hotel. I already had my "written" Buddhist pass from the abbot in Saigon. I checked out the action at Can Tho and Soc Trang in the South, both surrounded by Viet Cong. I flew on a couple of chopper missions behind enemy lines and another one to drop off medical supplies to an enigmatic Chinese guy who controlled an island in the

Mekong and kept it independent both of surrounding Viet Cong and Saigon. Lots went on in Vietnam that the media never covered.

To complete my studies of this extraordinary empire, I moved to the north to finish looking at all twenty-three of the South Vietnamese provinces, including Hue, the ancient capital, and frontier Ban Me Thout via airlifts on the versatile Caribou aircraft. The Caribous were well liked because they could descend to remote airstrips by such a precipitous nose-down drop that Viet Cong machine gunners in the jungle had a hard time hitting them.

By then the civil war scene had grown familiar to me, and so had the rules that the various power groups played by, so accepting Marie's invitation was not totally naïve. I knew how to thread my own way fairly safely through the hills lying between Dalat and the hospital. Marie had not only become a skilled nurse's helper, but also an expert in weaponry, including grenade rifles from the coaching of her well-armed visitors. Marie is one of the most fearless and capable planetary explorers I have ever met. I could write pages about her skill and ability to move at ease in all cultures and biomes. We got married a couple of months later with the neighboring Special Forces group and the hospital people attending. Looking carefully at the document before signing it, I found Vietnamese women enjoyed equal rights in marriage and had ever since the Truong Sisters had led a successful resistance against Chinese invasion centuries ago.

Newlyweds, John and Marie, in Dalat, Vietnam.

Jim Turpin needed engineering help for his hospital, to build outhouse toilets that cantilevered out over a steep canyon face to dispose of feces and urine. After I did that, he asked me to stay as his management advisor for Project Concern. Jim refers to this in his book, *Vietnam Doctor*.[2] I found Jim to be an intensely competent and resourceful, highly idealistic, "brotherhood of Man, fatherhood of God" inspired man. I took to this tribal but guerrilla-ridden Vietnamese jungle near the Dam Pao River in Dalat province, right next to the Ho Chi Minh Trail. I was one of the handful of individuals of any culture freely roaming about in one of the most fantastic countries and revolutionary situations conceivable, frequently landing between rocks and hard places that required all my ingenuity plus luck. This Project Concern experience taught me a lot about running small, highly technical, adventurous projects in difficult circumstances.

In 1966, while consulting to a gold and palladium outfit in the Cerrillos Hills near Santa Fe, New Mexico (whose flamboyant owner combined work with Interpol checking out promoters with promoting his own mining claims), I identified a nearby ranch as a high-energy, life giving place. In the spring of 1969, Marie Harding put all her capital, $14,500, into a down payment on the last quarter section of that small ranch while I raised $7,000 in capital from several partners, also anteing up my last $500 to buy tools to get materials (by taking down old houses doomed to fall under a "developer's" bulldozers) to build housing, animal areas, and carpentry, pottery, ironwork, and architectural shops. We started our project adventures.

Cerrillos, our bioregion, presented us with a very interesting combination of geology, ecosystems, and cultures. A transition zone between the Chihuahuan desert, high plains grasslands, and piñon and juniper biomes, where all these life forms mingle and compete, the Cerrillos Hills is a place where only the sturdiest individuals had survived. Culturally, the Cerrillos Hills had been an old mining district, perhaps the oldest in the United States, its turquoise having been highly desired by the Aztecs whose empire extended to Mount Chalchihuitl, a mile and a half from the ranch. The northwestern edge of its juniper-and-piñon-studded slopes marked the boundary of the Santo Domingo reservation; its hills had filled with mining claims and a few value-laden mines, its slopes been staked out by several ranches. The Hispanic culture interacted with the ranches and mines.

Synergia Ranch being built by its future occupants.

Occasional Indian hunting parties sought its rabbits and some of its few deer.

The Cerrillos area fell into the ambit of Santa Fe's complex life after most of the town of Cerrillos was washed away by flash floods caused by miners cutting down too many trees for their small smelters and most of the mines had been exhausted. Los Alamos and Sandia Laboratories, each located an hour away, ensured access to the cutting edge of the scientific milieu. Santa Fe also has served for generations as a favorite hideout for outstanding anthropologists and as an inspiration for certain artists and writers.

Marie and I named our find Synergia Ranch and found two other co-founders, Kathelin Hoffman Gray, theater director, and Bill Dempster, systems engineer, who are still operating it with us. I envisioned it as a complex, dynamic, adaptable system with many component parts, all interacting in a self-organizing (non-planned) way.

Marie Harding at her painter's studio.

Kathelin Hoffman Gray. Dance teacher, then director, Theater of All Possibilities.

Judy "Chili" Hawes. Versatile cofounder of Institute of Ecotechnics. Now director of October Gallery, London.

Our creative group soon included other key friends and fellow explorers Judy "Chili" Hawes, Mark Nelson, Robert Hahn, and Margaret Augustine. We figured Synergia Ranch's several vectors would produce surprising additional properties – this something new and additional is the gift emerging from the process called *synergy*. I first encountered the word "synergy" in my mining and metallurgical studies. In physical chemistry, synergy means, for example, that you can get stainless steel by putting together metals from whose separate properties you could never predict this amazing corrosion-resistant, beautiful result. My notion was that a synergy of our enterprises could build anything that our projects might require; these eventually included an ocean-going research ship, the *Heraclitus*, the October Gallery in London, and Biosphere 2.

At the ranch, various entrepreneurs set up various studios; Marie in pottery, along with her painter's studio and her horses; the brilliant young dancer and director, Kathelin Hoffman Gray, in theater; the dancer-musician Celia Davis in a clothing-costume shop; the versatile Joel Isaacson in metal works; avant-garde dancer Annette Longuevan in leatherworks; Robert Hahn in the laboratory making our cure for plant ailments; Phil Hawes (AIA) in plumbing; and Chili Hawes (a key advocate of natural childbirth in the Santa Fe community and active member of the La Leche League which encouraged breast-feeding) in health works. Bill Dempster, a mathematician and physics graduate from Berkeley who had been working at Lawrence Radiation Laboratories, took primary responsibility for the engineering systems, and quickly became a versatile actor and my long-term merciless antagonist in our rapid games of *Go*. Dick Brown, a master technician, and an artist whom I had met on the *Hrvatska* when sailing for Tangiers, co-founded the woodworks shop with me.

This profitable and productive woodshop, at its height run by four partners including Mark Nelson and Argos MacCallum, played a vital role in building our ranch facilities and furniture, its doors and windows, and also provided cash flow for the owners from the sales of hundreds of our unique tables, doors, and chairs.

Mark Nelson, a nervy, insouciant, and witty summa cum laude in philosophy and science from Dartmouth College, arrived shortly after startup, fresh from working at Time-Life Books and driving a taxi in Manhattan until threatened with mayhem by a passenger. After several years training in planting trees, he and I designed a five-acre sustainable orchard of apples, pears, peaches, plums, and cherries that we planted at Synergia Ranch. This orchard now produces our fruit and firewood, and provides camping sites and a strolling and picnic area. Mark has been my partner in crime (of freethinking) and biospheric theory ever since. Today he is recognized around the planet as a leading scientist in materially closed-ecological life-support systems and wastewater recycling. He and I co-authored the book *Space Biospheres*[3], which gave the original outline for the Biosphere 2 project. He got his PhD in Environmental Engineering from the University of Florida after Biosphere 2.

Mark Nelson in Western Australia, late '70s, starting up the Birdwood Downs savannah project; Chairman, Institute of Ecotechnics.

Margaret Augustine, in charge of building the Heraclitus *and designed many of our projects.*

Mark grew up in Brooklyn and Queens in New York City, a born logician, scrapper, and conversationalist. He and I created our bond through adventures in philosophy and worldviews, but also through fifteen years of hand clearing with adzes nearly a thousand acres of Australian savannah from invasive scrub, and then maintaining that clearing. There's nothing like three hours every morning chopping wattle to purge all poisons out of your system and clear the eyes. There are only a certain number of nuts who can chop wattle and appreciate Marx and Hegel, Plato and Aristotle, Nietzsche and William James, Kant and Whitehead, Rumi and the Upanishads, Tantra and Tao, Sartre and Greil Marcus, Malinowski and Levi-Strauss, Niels Bohr and Vernadsky at the same time. These philosophic adventures have their own excitement and danger, their own clearing of invasive scrub, their own poison and their own necessity for clear vision. Underlying our studies of these great

Bill Dempster, in charge of engineering and random element.

systems of thought and action was our sardonic appreciation of existentialist realities.

Besides a number of diverse enterprises run by the above individuals for their economic livelihood, people at the ranch participated to varying degrees in theater productions and the evolving line of work on ecotechnics. My special fields were basic building design, construction and theater. I set up our overall marketing system, called Biotechnic Bazaars, which became quite popular, and thus we financed the start of our fascinating and enlightening adventures. Each enterprise put in ten percent of its income to building and maintaining the infrastructure. We governed the ranch on Wilhelm Reich's "work democracy" principle: you got to speak on any area where you did responsible work. No one could pop off about something where they did no work, and no one could be excluded from decisions in areas in which they worked.

Marie Harding and I had started a company in 1967 in the Haight-Ashbury period of San Francisco called the *Enterprise for Developing Potentiality* (EDP). With $1,000 in capital, potentiality seemed a good direction to go! EDP operated somewhat like a private credit union, advancing small amounts (up to $5,000 eventually) to entrepreneurs who needed start-up equipment or materials. Intellectual capital gained about enterprise, ecology, and technics was open to all, laid out for critique and digestion in discussions at the dinner table and in special creative groups that studied particular problems in detail. By building our own habitations and shop buildings, no rent was required for the work premises, and by building and organizing a large scale kitchen, dining hall, storage rooms, and conference table in front of a log fire, as well as what amounted to a buying co-operative, monthly costs for food and utilities came to a quite affordable $45 per person.

The first EDP project in San Francisco, California, 1967. LEFT TO RIGHT: *Mike Ray, Marie, and John. Neal Cassady, among others, had coffee there.*

For agricultural experiments, after major discussion with two remarkable local innovators, Steve Baer and Steve Durkee, on growing food cheaply year round in high cold New Mexico, I constructed two semi-enclosed "grow-holes." Besides the cheap good vegetables, I tested hydroponic methods against soil-based methods of growing and found soil by far the best and most economical for complex, sustainable, recyclable agriculture. If they are properly worked, soils can increase their productivity. This work on soils and temperature, commenced in 1970 (and which continues today at Synergia Ranch), was to provide guidelines for the extraordinarily productive agriculture[4] in Biosphere 2.

We ran the kitchen at the ranch on the efficient style of a kibbutz near the Lake of Galilee at which I worked for a brief time in late 1968. I called it Synesthesia because we all agreed that the entire, differentiated, aesthetic continuum was needed to complete the protein, carbohydrate, fats, and vitamin calculations. Each dinner, therefore, had not only a chef or chefs, but also a person who did the environment, or ambiance; each kitchen team, rotated daily, put on a skit based on our theater exercises that aimed to highlight the mood of the day and increase the actors' ability. We treated each dinner as a feast, each Sunday night as special, and each equinox and solstice as a celebration of cosmic-solar-geo-bio-ethno harmony.

Nobody was ever late for dinner after a morning of adobe construction and afternoon of theater.

Convinced that no enterprise can succeed without its players being able to speak well, we made Sunday night dinners into "free speech nights," which constituted part of the training for our actors and built vital entrepreneurial skills. Each ranch member made a three to five minute speech on a self-chosen topic. No one could argue about what anyone said; it had to be listened to. Some plunged us into deep thought, some cracked us up in laughter, some produced silent but effective audience feedback that resulted in improved performance.

Left to right: Hank Truby, Kathelin Hoffman and William Burroughs at Synergia Ranch.

The ranch's planetary scope interested a cadre of creative individuals who became friends and in some cases colleagues. Among hundreds of people who helped us on our way, we met Konrad Lorenz and my old friend, Buckminster (Bucky) Fuller, on our theater tours and biospheric expeditions; William Burroughs, Charles Mingus, and Peter Fonda visited Synergia Ranch. We broke out the big blue porcelain coffee pot and gathered around the round table in the dining hall that always had "room for one more" when anyone interesting drove up the gravel road. Burroughs with Zen precision swatted flies that landed on the outside table. I asked Charles if he would give us the benefit of his philosophy. He said, "Tell me yours, and if I think you can understand it, I'll tell you mine."

Synergia Ranch served as a versatile headquarters to carry out our biospheric studies as well as a base for the theater company, the story of which will be told in the next chapter. Back on the ranch, four hours each weekday went to individual enterprises, two hours to theater during the week and much more on the weekend. And we all

The pottery at Synergia Ranch in the early 1970s.

Pottery was produced from clay discovered nearby.

spent four hours a day in good weather, which is most of the time in the Cerrillos Hills, building – by hand, wheelbarrow, hammer, cement mixer, shovel, electric drill, pliers, adze, saw, and scraper – our required buildings and shops from old timber, Forest Service-thinned pine trees, and the adobe that lay underfoot, which we formed into six-inch thick Pueblo-style bricks, heavy and strong, with the right mix of sand and straw, water and sun. That adobe deposit was my most important mining discovery! It must have saved us at least a hundred thousand 1969 dollars that we didn't have. Bill Dempster organized a high-morale crew that tore down old houses in town that faced that meaningless destruction called development. Fine aged lumber went to the woodshop to become ceilings, window frames – whatever we needed.

Synergia Ranch provided my colleagues and me with a permanent base for economical, healthy, and all-around living, including horse riding, barbecues, and great hiking. We also established a library and a chemical laboratory for research into soils and plant growth. Robert Hahn and I came up with "Uncle Bob's Miracle Root Cure," which

Pottery and furniture design and production were major enterprises at Synergia Ranch.

Synopco drawing of Project Llano site on Palace Avenue, Santa Fe, New Mexico.

Argos MacCallum, actor, poet, and craftsman making furniture in the woodshop at the ranch in the mid-seventies.

sold well. The idea was that it contained all the basic supplements a plant could need (like a good multi-vitamin pill). We carried out grow-hole experiments on the interaction of plants and soils and also developed architectural and technical design skills that would later create Biosphere 2. To finance our projects, after first making crafted products to sell at our Biotechnic Bazaars, we formed a construction and design company, Synopco (Synergetic Operations Company), which designed and built a total of thirty-seven fine adobe homes in Santa Fe, thirty-one of those in the now-much-appreciated-in-value Llano Compound on Palace Avenue. This project, together with the ranch-crafted windows, doors, and furniture that went into the houses, capitalized most of our work till 1978. The ranch charged Synopco five percent of its sales for serving as its logistic and office base. Synopco was the forerunner of the present Global Ecotechnics Corporation.

At one time, ninety people worked on our Synopco projects in Santa Fe on my four-day work week for four hours a day at four dollars an hour ($2.15 was the national/state minimum wage at the time). That meant intelligentsia could afford to work for Synopco while still having time for their main tasks in writing, arts, and activism, and meet-

We worked with the county to rewrite the building code to allow two story adobe homes.

ing all their living costs as well as interesting people. The apprentices, many of whom became genuine craftsmen, included women liberationists eagerly learning practical skills previously reserved for males, Chicanos engaged in the land rights movement centered on Tierra Amarillo, and Indians engaged in their rights movement. A few freethinking rednecks drifted up from Texas.

Synopco thrived, a fun outfit with a perfect safety record, probably because no one got overly tired from four hours of work a day. Kathelin Hoffman was Synopco's ingenious president and chairman for five years; she worked with Chicano subcontractors, local artisans and helped the first women in the state get their electrical, plumbing, and building licenses. Understandably, we experienced some opposition from big developers with their fake adobe. All of this gave our actors valuable experience in high intensity situations and developed our ecotechnic systems approach.

Crew at Project Llano.

Ed Bass, a Yale graduate, from Fort Worth, Texas, was a lean, ponytailed Navajo rug trader when we first met him. He approached Synopco in 1973 to finish construction on a million dollar house he had started with his construction company, Badlands Conspiracy. We took and finished that contract, and then entered into a joint venture with his company, Fine Line, Inc. to build Project Llano in Santa Fe, which took until 1978 to complete.

Margaret Augustine, who became the leader of our architectural efforts and inspired project director, showed up at the ranch in 1973 at age twenty-one with her boundless intellect and unconquerable spirit, fresh from three years adventuring on her own around the world, while earning her living by ingenuity and sewing skills. She quickly apprenticed with me on the study of historic architectural forms that provided inspirations for Biosphere 2 and on drafting and design to architect Phil Hawes, who had trained under Bruce Goff, and joined our efforts in an on-and-off way.

Margaret Augustine working on Project Llano design.

Off for Phil meant off on his own in some personal adventure, like walking by himself from the Fitzroy to the Timor Sea through the wilds of the Northwest Kimberley Plateau in Australia. His nickname, Thundercloud, came from the kick he obtained from allowing his plane to soar on updrafts into the center of those seven- and eight-mile-high cloud formations. *On* meant stopping his adventures for a while to work on some of our projects while exercising his magical drafting skills. He had studied with two great masters of architecture, Frank Lloyd Wright and Bruce Goff, and was building innovative adobe houses in Santa Fe when we met him. Phil worked for a while at Synopco, then on the Vajra Hotel project at the startup, then voyaged on the *Heraclitus*. I had met Phil in the spring of 1966 when living in a small compound in Cerrillos near the edge of the San Marcos Wash, riding horses, and consulting on mines. An outstanding college wrestler, Phil married the beautiful Chili, a Peach Queen of Colorado and scholar of Rimbaud and Baudelaire at the Sorbonne. Chili had a natural gift for management and now directs our international art gallery in London.

Phil "Thundercloud" Hawes, architect-adventurer.

Phil took Margaret on as his apprentice at our Synopco project in 1973. Margaret was lead designer for most of the highly prized houses at Project Llano, eventually becoming a licensed New Mexico architect. She started her other line of work, project management, by taking over and completing construction of our research vessel, the *Heraclitus*, from August 1974 to March 1975 in Oakland, California. Later, she personally coordinated my conceptual vision and detailed invention and construction specs for Biosphere 2 with the architectural design, approved by herself, Phil, Bill Dempster and myself. Also, as project director, later becoming CEO, she was responsible for the construction and operation of that complex system.

My great Tibetan friend, Paljor Thondup, of a royal line of Kham, who stayed a while at Synergia Ranch till he got the hang of American life, gave his Happy Buddha belly laugh when I described my reaction to people looking assiduously for a guru to relieve them of personal responsibilities. "Ha," Paljor grinned. "They come up to me on the sidewalk and say, 'You're a Tibetan, I want to be enlightened.'" Paljor was an expert horseman who fought his way out of Tibet in 1959 at the age of seventeen. His group had lost five hundred and seventy out of six hundred. The "crazy" Mahamudra master Chogyam Trungpa was another of this thirty that survived. A dead shot with a .22, Paljor could hit a jackrabbit between the eyes. "You come from the Wild West," he'd say to me, "and I come from the Wild East."

Paljor Thondup, early 1970s. Founded Project Tibet at Synergia Ranch.

Bill Dempster, a mathematician and systems engineer, Mark Nelson, a horticulturist and biospherics theorist, and I kicked things around at a profound level of theory and a complex level of practice without wasting time on conversational niceties. Kathelin Hoffman joined us in developing the new discipline I called *ecotechnics*, the ecology of technics and the technics of ecology being its two main branches. She made sure we didn't leave theater and psychotechnics out of the ecology.

Crew working for Synopco construction company that built Project Llano on a four day week, four dollar an hour, four hour a day schedule.

Kathelin Hoffman, President of Synopco: the Theater of Action.

The *ecology of technics* means evaluating the total system of inputs and outputs from and to the biosphere of any technical system; understanding the *technics of ecology* was paramount to producing a sealed system within which another biosphere could live. The technics developed by observing and measuring that class of complex systems can be directly applied to study natural systems and to monitor large-scale projects. By 1973, Bill, Mark, Kathelin, and I refined the concept and practice of ecotechnics to the point where we, together with Chili Hawes, founded the non-profit Institute of Ecotechnics in New Mexico. Mark still chairs our now forty-odd members and entrances us with his epigrams when introducing speakers at Institute of Ecotechnics conferences, as he has except for 1975-81, when he became fully involved in creating the Synergia orchard and in setting up our five thousand-acre savannah systems project in Australia. Chili remains the deft Treasurer and Keeper of the Institute's office at Global Ecotechnics' Lundonia House off Queen Square in London.

In 1980 we integrated our architectural and design work into a new corporation called Biospheric Design, whose primary mission consisted of designing and building projects initiated by the Institute of Ecotechnics. Margaret was CEO and she, Phil Hawes, Marie Harding, Bill Dempster, Ed Bass, Kathelin Hoffman and Robert Hahn all bought stock to start up the new company. No one owned more than twenty percent. Biospheric Design became possible because we had completed our initial learning curve building Synergia Ranch and its various artisan shops and labs, followed by Synopco's Project Llano in Santa Fe.

The smaller forerunner of Synopco, Transflux, finished building two one-of-a-kind houses on Camino Manzano in 1973. I worked on the special windowpanes of one, carving out curved shapes in the glass with Dick Brown; Phil called the bathroom the Aphrodisium. We used railroad ties and telephone poles to set the adobe walls. After that, I located a large area on Palace Avenue about eight blocks up from the Palace of the Governors in central Santa Fe. This small llano (flat area) overlooked a steep-sided arroyo. It seemed an ideal place to reintroduce real adobe buildings into the central city area; that would

One of the adobe homes at Project Llano, Santa Fe, New Mexico.

Project Tibet, a center for the renaissance of Tibetan culture, was another Synopco construction project built near the Santa Fe Plaza in 1978. It continues to be a center for education about Tibetan culture.

strengthen the city's fabric against the suburban sprawl that rapacious developers were inflicting upon this magnificent traditional capital, founded before the Pilgrims landed on Plymouth Rock.

The Hispanic family that owned the llano, now comprising over three hundred individuals, had deliberately spread the ownership in order not to lose the land to a fast-talking Anglo. However, their leaders liked our style, as seen at Camino Manzano and Synergia Ranch, where we made thousands of adobe bricks by hand in the old wooden forms. They volunteered to get all of those relatives' signatures, thus enabling the land to be sold to us. At last, I could build the thick-walled adobe paradise I had dreamed of, with efficient plumbing and electricity, with the best trees preserved, the occasional fruit tree planted, and paths and certain areas to remain common property. Our Hispanic friends said, "It will improve Santa Fe for all of us."

Synergia Ranch was started in 1969 as a center for innovation. Here the theater dome has just been finished. Landscaping to follow.

In addition to the two *tour de force* houses on Camino Manzano and thirty-one houses at Llano Compound, we also built five new buildings to make the base for Project Tibet on Canyon Road, and Plaza Alegre on the road to Taos. The success of Synergia Ranch and the Santa Fe projects inspired and created the intellectual and technical capital for expanding into the more demanding type projects that would take the next step toward building Biosphere 2. Our toast: "Step by lucky step." Our motto: "We deliver."

Theater of All Possibilities

CHAPTER THREE

All the world's a stage,
And all the men and women merely players:
They have their exits and their entrances,
And one man in his time plays many parts …
SHAKESPEARE

You cannot step into the same river twice.
HERACLITUS

IN 1969, MY MIND, ALREADY PREPARED by Colorado Mines' Historical Geology, was dazzled, "blown-out" by life's awesome powers when I discovered and was enlightened by that classic scientific masterpiece with the metaphoric title, *The Ecological Theater and Evolutionary Play*,[5] by Evelyn Hutchinson, an outstanding ecologist. Evelyn had been enlightened by the work of Vladimir Vernadsky, the magnificent Russian scientist and founder of biospherics. Vernadsky's son, George, had escaped the Soviet Union to teach at Yale as a colleague of Hutchinson's and had turned him on. One of Hutchinson's students, Howard Odum, later became one of my key scientific friends and colleagues at Biosphere 2. Hutchinson showed me how to bring together two of my passions, the Biosphere and the Theater. I could now integrate my enterprise.

Maybe, *The Ecological Theater Enterprise*, or *The Enterprise Play of Ecological Theater*. It finally emerged as *Space Biospheres Ventures* when the time came to act.

Not for nothing had I spent years studying theater, since my mother spent her last few extra dollars to take me, at age ten, to see that wonderful virtuoso actor Maurice Evans play Bernard Shaw's *Man and Superman*, including the masterpiece scene in hell. Spellbound by

Vladimir Vernadsky, great biogeochemist, founder of biospherics, a genuine hero of thought and action.

IMAGE ON LEFT: *Allen circa 1963, exploring for coal deposits in northern Arizona for Lilienthal's Resource Development Corporation.*

Evans' acting, I watched the power of free thought, fearless emotion, trained muscles, and uncanny attunement to people's attention transform a man for more than two hours into a Superman. Evans showed me that I could have as much or more fun playing with physiological transformations as with the physical transformations in my chemical lab on the back porch.

Evans and Shaw produced the same level of liberating influence on my intellect and emotions as the chemistry set I had managed to buy for ten dollars out of the money I made out of my weekly two page newspaper. *The Chit-Chat* had one page for local news (such as where any new potholes could be found and who had whom over to dinner) and one for foreign news (where I kept my subscribers up on the intricacies of the Allied cause because the local paper was isolationist). I had started the paper in 1940, selling it to my forty or fifty customers for two cents (when a hamburger was ten cents). I immediately put the chemistry set to use, stripping elements out of molecules and making molecules out of elements and turning one molecule into another. It was years later before I got to work on projects changing forms of an atom (making zircaloy rods to control reaction rates in Admiral Rickover's nuclear submarines).

Wow! Experimenting with those magical and dangerous materials – phosphorus, mercury, sulfuric, nitric, and hydrochloric acids – and observing their forms, pH, heat, composition and state transform according to definite principles, I realized the power of science. I was captivated by the lure of infinite secrets waiting to be discovered in a universe of awesome energies and ingenious, enterprising molecules and atoms. Evans showed me the power of experiential science, whose transformational displays are called *art*. Shaw's text on atoms of human consciousness, together with Mendeleyev's text, the periodic table of physical atoms, introduced me to the gigantic physiological substructure that underlies the human mind. My attention could focus power in my brain the same way my grip could focus strength in my hand. Later, I would endeavor to make experience-ments with biospherics, integrating the powers of art, science and enterprise. Hutchinson's book showed he had made this integration; what had I been waiting for? The Biosphere 2 project began in my mind at that moment.

Before reading Hutchinson's bombshell, I could barely hazard a guess as to how to live in this marvelous world I had glimpsed, inhabited by these demigods of theater and science who survived the determined stalking of villains and the remorseless reactions of elements. I knew I could get there if I was lucky enough and worked as hard as possi-ble. My secret *mantram* came to me: *What man can do, Johnny can do.* I never told anyone my secret. I kept my silence and my reward was a certain kind of precision. I didn't know the word mantram then, but I did know the word secret. Whenever I found the going too tough, the secret magical phrase emerged as if of its own volition. I never intentionally said it aloud; it would have been tempting fate.

But when the secret spoke in the midst of almost giving up, somehow new strength poured into me from what seemed an inexhaustible source. Perhaps this confidence came from actually having seen the human accomplishments performed by the likes of my grandfather, my grandmother, Uncle Bus, and Cousin Buck.

My multi-leveled life back in Manhattan in 1962-63 saw me zooming around the world techno-theater that stretched from Iran and Liberia to the coal deposits of the Energy Empire in Western United States and Alaska at one pole, to Joe Cino's no-holds-barred Off-Off Broadway theater on Cornelia Street at the other pole. Joe let me have the front table while I drank my coffee so I could watch as actors did their thing about three to four feet in front me. I grokked every muscular twitch, popping out glint of sweat, and talking eyeball. Zip-ping around trying to locate pay dirt in this magnetic field of On-

On History and Off-Off Art, I followed up any traces of gold veins to discover the mother lode from which planetary theater had created seven different kinds of basic dramas or conflicts: tragedy, comedy, social, heroic, epic, absurd, and mystery. The mother lode, Stanislavsky informed me in his *An Actor Prepares*, was: *Magic If*.

Kathelin Hoffman, my partner since 1967 in all things theater, was a top student of that most original and ageless dancer, Anna Halprin. She possesses a ruthless, brilliant sense of style. Together with Marie Harding and my old philosopher friend, Ben Epperson, we developed our Biospheric-Ethnospheric Theater and called it the Theater of All Possibilities (TAP). A Sufi asked me, "Do you really want ALL possibilities?" Well, certainly not, except in the world of *Magic If*. But, yes, I do, in that world.

Kathelin and John working on a script in London.

The four of us, with a six-foot-four actor adventurer named Mike Ray, a daring Mormon girl named Susan, and two others, began TAP in 1967 in San Francisco in an old two-story Victorian house on Sutter Street. The theater quickly grew to twenty actors and we soon had to rent an abandoned grocery store on the corner for acting classes and rehearsals. We decided to base TAP's plays on no single culture, but to take themes that "sailed with a contemporary wind," as Brecht demanded, from any part of the ethnosphere. Drama, at its best, integrates the conflicting memes, scenes, themes, and dreams of a culture; conflicts that find a cathartic understanding in *Magic If* metaphors, but result in bloody endings when acted out in *If – Then* algorithms.

Kathelin and I eventually decided that the fundamental theme of our theater would be William Burroughs' call for artists to explore human intentions towards the planet, using a basic style we called Organic Realism. Organic – based on each actor's proprioceptive being-perceptions; and Realism – based on each actor's character name and its cultural and economic-political perceptions. We saw

Some actors in the Theater of All Possibilities ensemble, 1981.

the human body-mind as the biggest unknown factor in the universe and human thought (or mind-body) as the most unconscious factor in consciousness. We regarded theater's role as "the revelation of the inner life of humans." Soundly basing the theater on Organic Realism allowed us a grand spectrum of styles from which to choose to produce special plays: Romantic Realism, Sur-Realism, Grotesque Realism (from grotto as the symbol for the subconscious), Hallucinatory Realism, and Satirical Realism; "Organic" is what put our twist on these "tiger tails" of reality.

Our general procedure was that I, as dramaturge, would write new plays and make adaptation-translations of classics in order to bring our team to an intimate understanding of ethnospheric forces, present and past. Kathelin made illuminating critiques. I taught the acting

IMAGE ABOVE: *The TAP bus prepares to head out on tour.*

BOTTOM: *TAP ensemble on the road.*

classes, while Kathelin taught dance and Marie taught sound. We aimed to perform on all seven continents, and in all types of cultures. With a production in Antarctica in 1989, TAP reached the first goal, and we have played in most types of cultures.

In one year (1974), I finished two new plays using many passages from the authors portrayed – *Brecht and Artaud* and *Shaxpere and Fitton*. Kathelin directed productions in Berkeley, San Francisco and Los Angeles, while we were building our oceanic research vessel, the *Heraclitus*. *Brecht and Artaud* examined the artistic life and meaning of two of our six basic influences in forming the Theater of All Possibilities (the other four were Stanislavsky, Balinese, Japanese theater, and Guerrilla theater). The play's "magic if" was that these two great artists met in 1933, just before Hitler came to power. The French could have had their dada and the Germans their emotional flagellations on stage rather than on the battlefield.

Would Artaud have helped the Germans experience catharsis of their *sturm und drang*? Would Brecht have helped the French distance themselves from their dreams of glory? For *Shaxpere and Fitton*, I took Frank Harris's vision of the beautiful lady-in-waiting, Fitton, as being the dark lady and dark force of Shaxpere's sonnets. I used the progression of those sonnets as the through-line of the play. Both were received well by UCLA and Bay Area audiences and reviewed by the *Berkeley Barb*.

With our ecotechnic studies and projects, we still devoted two months a year to touring. Our motto on tours: *Create and run*. That meant playing for one week only or sometimes for one performance in a theater. Two weeks would be exceptional, as in Paris, or Berlin, or London. We set those time limits because critics seldom reviewed a play they didn't think would last at least three weeks. We espoused (naively, perhaps) the ancient notion of the anonymity of the artist.

We designed theater tour routes to synergize with gaining bio-techno-ethnospheric knowledge and for project knowhow, and called them "ecotours." For example, a portion of one tour of France included camping out at Carnac's ancient menhirs, and descending beneath the Cathedral of Chartres to sacred places. During one of our three visits to Mont St. Michel, we stopped at the dank tarns of Auber and ghoul-haunted woodlands of Weir in the Massif Central. We studied water uses in the Industrial Ruhr while playing in the Werk-statte at Düsseldorf. We cooked our own meals while traveling on both our American and European buses, read from our select library, and, in a pinch, slouched in our seats to sleep, sometimes pulling off into inviting woodland, sometimes loafing in a relatively plush apartment or two given by the theater we played at or its friends. We loaded the bus top's steel frame with costumes, sewing machine, props and painted scrims. Our tours usually involved fourteen actors who also did lighting, make-up, costumes, and so on; we managed to cover our expenses, sometimes by living off the land, thanks to friends who allowed us to pick cabbages or apples.

An ecotour in Samarra, Iraq, 1977: a two week adventure from Nineveh to Babylon studying the fall of ancient civilizations due to siltation, saltation, devastation, etc.

On our first tour of America east of the Mississippi (in 1970, while on our way to perform at Kingston Mines and Body Politic Theater in Chicago, and before proceeding on to theaters in Atlanta, Georgia and in Coconut Grove, Florida), we stopped at Carbondale, Illinois so I could discuss the design of our new fifty-foot diameter theater dome at Synergia Ranch with Buckminster Fuller. The master of dome design, Bucky believed in beautiful and efficient architecture and lived in one of his domes himself. He asked me if I wanted to see his famous World Game that he had presented in dozens of countries and I said, "Sure." He grinned owlishly and took me into his bedroom, pulled out a drawer and handed me a few scraps of scribbled paper. I took this gnomic behavior to mean that the World Game existed in his head and that he changed the presentation to every audience. I grinned back. A deep bonding formed.

Bucky Fuller at the Institute of Ecotechnics Galactic Conference, 1982.

Construction of the geodesic dome at Synergia Ranch.

I had first met Bucky Fuller at Harvard, when he spoke before several hundred enthusiastic Harvard and Radcliffe men and women. Even with their famous capacity for attention, I wondered how many would make it to the far end of his speech. Fuller's being-challenging technique in speechmaking consisted in starting anew on his comprehensive thought at every major talk to see if he would come to the same conclusions and, if not, adopt the new position that had evolved. His technique had basically the same goal as Konrad Lorenz's: to come up with something new and better each time you went into action, without losing any of the good stuff. Anyhow, by the end of four hours, the audience had dwindled to a handful. After another thirty minutes, there were only four of us, hungrier for ideas than food. "Now we can get down to something," Bucky said, with his contagious grin, descending from the platform to stand around the deserted coffee table with us.

Buckminster Fuller was a real metaphysician. He represented in the flesh the best in my culture's transcendental lineage – Thomas Jefferson, Ralph Waldo Emerson, Henry Thoreau, Herman Melville, Walt Whitman, Willard Gibbs, William James, Frank Lloyd Wright. Self-Reliance. Envy is folly. Imitation is suicide. Simplicity. Simplicity. Simplicity. I wear my hat as I please indoors and out. Free Energy. The varieties of religious experience. Synergetics. Marie and I named our ranch Synergia to honor this ever fresh, ever practical worldview, a worldview weighed down by no encumbering paraphernalia; a worldview with no predetermined endpoint. Loaded with confidence, skill, and room enough for a free spirit, it was the most hidden worldview on the planet because it was open to all. No costumes, no drill, no handed-down texts forbidden to be analyzed, no authority figures. "Henry, what are you doing in that jail?" "Ralph, what are you doing outside this jail?" Find a koan or Mullah story to beat that.

TAP performance of an original play, The Energy Empire, *at Synergia Ranch.*

Bucky advised that for the best theatrical acoustics I ought to make a five-eighths dome and not a half-dome. From the first, Bill and I had been very interested in using Fullerian synergetic geometry in any buildings designed for complex activities such as a biosphere. Under Bill's rigorous supervision, we built the dome at the ranch in 1971 to his computer generated graphic design and personal quality control of Bucky's and my specs. Bill also had some great ideas on how to improve the water tightness of the doors to the dome. We used canvas for the first dome covering, thus checking out the sail material for the ship I was already talking about. We built the ranch showers with ferro-cement walls to test that material's interaction with water. I figured if it would keep water in a shower, it would keep water out of the ship.

Legends of the Heraclitus

CHAPTER FOUR

Swell after swell of ocean rapture lifted me up
and dropped me down,
Heraclitus and all metaphysics appearing disappearing
under the blue inner sky,
I even forgot to fear that the moment would end and it never has.
JOHNNY DOLPHIN

THE VISION OF A SHIP that could cruise the world oceans with special attention to coral reefs, estuaries, and big rainforest rivers arose
in me in 1965 after living with Sea People on a Chinese junk located
just off Kowloon. At the time, I was the engineer and management
consultant for Project Concern, which, besides its Montagnard Hospital in the mountains of Vietnam by the Dam Pao River, ran three
hospitals in refugee-filled Hong Kong.

Jim Turpin, Marie Harding and I left Vietnam soon after President
Johnson violated his "campaign oratory" and escalated the war by
bombing North Vietnam. General Giap, in return, escalated the frontier region fighting. Our Dam Pao River site sat on that inflamed frontier, near the Ho Chi Minh Trail; that section was guarded by a handful
of Special Forces. We flew out of Vietnam on an Air Force plane to
Hong Kong to investigate and work at Jim's projects there. One project
was secluded in the Walled City, neither British nor Red, where we
were the only "round eyes" that I saw. Another project nestled among
the refugees clinging to the steep hillsides in shacks. And the third,
Jim's medical center, headquartered on a spacious, polished junk
amongst the Sea People, where we lived in the midst of this cheerful,
productive clinic. We came to know these Sea People very well. I read
how the great Admiral Zheng He, in the fifteenth century, European
calendar, had sailed junks all over the Indian Ocean, a great part of the
Pacific Ocean, and had probably sent a fleet around the world.

IMAGE ON LEFT: *The*
Heraclitus *in Phuket,*
Thailand, 2007.

These Sea People opened my eyes to how harmoniously people could live together. Gossip claimed that some of these Sea People would never go to shore. They certainly didn't have to, they had Wedding ships, Ice ships, Equipment ships, Residence ships, and Party ships. I had experienced nothing like it anywhere in the Western, Hindu, Islamic, or even Confucian cultures. Only self-governing tribal peoples came close to matching this ever-alert, ever-cooperative, ever-cheerful way of life. This rich and happy experience of living in such a skilled and closely-knit community, in constant touch with the ever-changing, sometimes dangerous elements, gave me a whole new way of looking at how to live on and enjoy the sea. The variegated moods of the sea had to be given a lot of credit for creating the Sea People's instant adaptability. So, conceiving and building a ship based on the classic Chinese junk celebrated and expanded upon that experience.

In 1973, I saw that the next step toward creating an artificial biosphere was for our creative group to master highly technical design and construction skills. Without a ship, I didn't see how we could understand biospherics. Two-thirds of the atmospheric surface of the biosphere touches ocean, and without the water cycle driven by sun and ocean, the rest of Earth would be desert. Without the tides driven by sun and moon, the oceans would lose much of their diversity.

To understand our biosphere, one must understand the oceanic biomes, winds and currents, and their tidal, geological, climatic, and evolutionary interactions with the terrestrial biomes. Understand the tropical convergence zone, the Gulf Stream, monsoons, the doldrums, the trade winds, the Roaring Forties, freak waves, the Antarctic banana belt, El Niño, estuaries, tidal bores, marshes, lagoons, coral reefs, bays, sounds, seas, channels, salt lakes, subduction, hot spots, ridges, and tropical storms, among other realities. Understand how the productivity of the ocean depends on animal diversity, and how the land's productivity depends on plant diversity. Understand

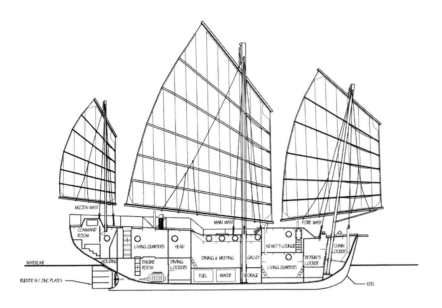

Allen's and Dempster's design for eighty-four foot sailing vessel: a modified Chinese junk built with ferro-cement hull and keel.

the difference between mucks and soils. We needed to build an ocean-going ship, which could deal with coral reefs, rivers, and Antarctic waters.

My notion of the crucial importance of understanding the ocean and Sea People to understand the biosphere pushed me to a critical point. If I went ahead with building the *Heraclitus*, everything I had done at Synergia Ranch could be tossed into turmoil. It meant tremendous sacrifices for all of us. It meant splitting the winning organization in Santa Fe into two groups, one doing Synopco and one designing and building a ship. Splitting up is always a risky proposition. We could lose it all.

I had grabbed hold of the tail of a tiger in wishing to understand the whole biospheric system in order to build a physical model of it. Building the ship, though it seemed a necessary step, would push me

Maiden voyage of the Heraclitus *out from Oakland, California, 1975, running into a gale. The crew quickly became seamen.*

Heraclitus *design board at Synergia Ranch.*

to the edge and cascade inescapable consequences upon my head and the heads of my closest friends and allies. Yet I knew that without a world-ranging ship that would expose me and the creative group to the vicissitudes and glories of the ocean, my talking about understanding biospherics and eventually building Biosphere 2 would become fakery. Just endless chatter.

I had to become a flexible being that directly experienced Planet Water or I would remain in personal unreality about the biosphere. I had taken passage across the Atlantic and Indian Oceans, down the Gulf of Siam and on the China Sea, taken three voyages on the Mediterranean, gone back and forth three times across the English Channel, but being a passenger I felt (correctly) that all that was like taking train rides across continents but never getting out. One night it all came together and I quickly sketched the conceptual design of a biospheric ship. If we could build it, crew it and cover operating costs, then we (the Institute of Ecotechnics) would possess the means to learn about the ways and life forms of the World Ocean.

John and crew at the Fifth Avenue Marina, Oakland, California.

So my life underwent another transformation when we began the design, building, and operation of the *Heraclitus*, the now legendary research vessel named for the least pretentious, yet most enigmatic philosopher. The command to do the *Heraclitus* definitely came from somewhere else and I experienced fear and trembling.

Fourteen of us left Synergia Ranch in Santa Fe, New Mexico to build the research vessel *Heraclitus* as volunteer labor. Chili Hawes stayed behind to run the Ranch.

Between August 1974 and March 1975, we worked up to fourteen hours a day at the Fifth Avenue Marina in Oakland, on the estuary just up from where Jack London had built his self-designed *Snark* to sail across the Pacific sixty-five years earlier. Bill Dempster finalized the design and supervised the engineering work on the ferro-cement hull. I visited the Schiffbauhaus in Hamburg; Bill contacted top engineers for help on fine points.

Margaret Augustine, now the Institute's apprentice architect, supervised the demanding construction, as well the laying out and sewing of the sails. With Robert Hahn, she also made the water tanks. Kathelin Hoffman did extraordinary work in obtaining materials either with high discount or by donation: making close friends with guys who ran supply houses and second hand warehouses. Marie Harding, who, as usual, could be found wherever troubles boiled up, ran our finances and became a champion at tying wire mesh. Robert Hahn won the nickname "Hands" for his technical skills. Mark Nelson slogged away like a Stakhanovite for a couple of weeks. The other seven also worked handsomely and played with high spirits, but did not go on to accompany me in building Biosphere 2. With the completion of both Synergia Ranch and *Heraclitus* under our belts, an *esprit de corps* emerged that I felt was truly elite, a sort of special forces unit working on ecological frontiers. "Ready for anything, we hope," became our motto.

We worked seven days a week, only taking a half day off Sundays, with the exception of a four-day theater junket with Kathelin's and my production of *Carneval of the Seven Sins*. We performed at UCLA, picked up $2,000 towards our living expenses and had a wonderful break from our high-pressure routine. The late great Dr. Roy Walford, who later designed and ran Biosphere 2's medical system, had been our sponsor. Roy had "awakened" during 1968, in Paris, after he had left his innovative but rules-bound lab to take part in the Situationist Street. Our living costs stayed low, thanks to Bill Dempster's mother who contributed his grandfather's impeccable three-story redwood house near the university for us to live and cook in during the first two-thirds of the construction.

The Junkman's Palace restaurant on Telegraph Avenue, Berkeley, California. Great scene; great food. Sold when the Heraclitus *sailed out.*

We covered most of our food expenses from the profits of a café, The Junkman's Palace, which I designed and located just down from Le Bateau Ivre, that Telegraph Avenue watering hole for the hip, the nubile, the intelligentsia, and the voyeur in Berkeley, California. Buying our crew food together with the café enabled us to get wholesale prices, and gave us a cool place to meet other than at the ship site. We marketed our specialty creation, the "greatest hamburger in the world," under the name Berlin Burger. We found a reliable ground beef supplier and a source for excellent bacon, two strips of which lay across the patty, along with crisp lettuce, juicy onion slices, two slices of a tasty sweet pickle, a thin slice of a beefsteak tomato, and a special secret sauce. Another Junkman's specialty was our homemade pie, especially pumpkin, a splendid recipe that emerged from pie-making contests at Synergia Ranch. One evening, the San Francisco Symphony Orchestra reserved the Palace for dinner.

We had our share of rough times, notably, when the hull of the *Heraclitus* already stood high in the air and accumulating problems made finishing the complex work seem impossible. Gloom and doom dominated our talk at breakfast and dinner. Simple truths showed that we should return to the ranch. We definitely had profitable, beneficial, and beautiful things to do, like building more adobe villages in Santa Fe, doing theater tours of Paris, Berlin, Rome, and New York City, and holding ecotechnic conferences. I could finish my novel, *Thirty-Nine Blows on a Gone Trumpet*, and even write a new black comedy or two satirizing the Triumph of Reaction in America. I could ride a horse around my beloved Cerrillos Hills on country that didn't yet have a trail.

I always championed learning by doing, but maybe we *had* bitten off more than the toughest jaws could chew. Even if we finished the ship's structure, launching it would be a delicate operation and sailing it around the world on a shoestring could turn into a nightmare. Bill Dempster calculated that we could only safely launch the ship during a monthly four-hour window, at high tide, and then only after jackhammering out a channel through the concrete ruins of some previous dreamer's project, lining it with sandbags to create a "birth canal."

I could find no error in Bill's calculations. We would have to install the masts, test out the engine, train a crew, pass Coast Guard inspection, and somehow find the money to stock the *Heraclitus* with food and equipment. Then the *Heraclitus* would have to sail nonstop to Hawaii, since that was the first island on the way to the golden seas of Southeast Asia, where we planned to make our first studies of the oceans and its related cultures. To get to Hawaii we would have to master the mysteries of wind and current. Bill began taking the Coast Guard's Piloting and Navigation classes.

Where would the money come from to run the *Heraclitus* if we did finish it? Even though we could all "make the eagle scream" by clutching each dollar so tightly, where would the people come from to replace those who had started the venture? Would the ship eventually simply have to be abandoned on a far off beach, like Jack London's *Snark*? How would rescuers get there if worse came to

Making the sturdy ferro-cement hull.

Ship construction site shortly before launch, Fifth Avenue Marina, Oakland, California.

worst? It would take a generation to create the nucleus of a Sea People able to go anywhere and live well and in harmony with the World Ocean. Who and what could we count on?

We drove away from Mrs. Dempster's classic redwood house in our yellow theater bus, each sunk in somber silence. Suddenly, I saw the magic shape of the *Heraclitus* hull rising boldly against the sky. Something grabbed hold of me and spoke for the first time. "Stop the bus! Let's take a last look!" this voice commanded. The bus pulled over and stopped.

The voice that came out of me said: "We have left the *Heraclitus*. What we started to do was, and is, impossible, as we all agree. However, we are no longer the same people. We are a group of highly skilled actors and ecotechnicians who face a secure future in New Mexico when we return from our wonderful experience in Golden California. We are at a moment of freedom. And what do we see in front of our eyes? A fully designed, half finished ship around which lies nearly all the material needed to finish it; plus a rent-free site and a rent-free place to live, located in one of the most beautiful and stimulating environments in the world."

We moved into an abandoned tin shed near the hull of the *Heraclitus*. My bunk space was about four feet by eight feet of concrete floor. The separations for privacy ranged from plywood sheets to thin boxes stacked on edge. Marie and I had started a small underground film company, Flash Films, and she recorded some highlights of those days. We successfully built and launched a two-decked ship: eighty-four feet in length, with a beam of twenty feet and a draft of seven-and-a-half feet. That launching marked the Institute of Ecotechnics' entry into the world of complex dynamic systems design, construction, and operation.

The *Heraclitus* had to be comfortable enough for living and working in port, at sea, tied to a tree on the bank of a tropical river or anchored in a coral lagoon, since we had no money to pay for accommodations ashore. The library on the ship contained enough books on ocean geology, ecology, biology, and ethnology, as well as history, philosophy and avant-garde novels, so that Captain Nemo wouldn't have to read anything twice even if he took an unbroken Twenty Thousand Leagues trip. The dining room was large enough for the crew to rehearse theater and to hold small conferences at remote ports and beaches. Eliminating tables and chairs by using floor cushions gave lots of space, and made visitors feel as at home as we did.

We made walnut tables embedded with turquoise that Chili Hawes and I had mined near Synergia Ranch and special chairs in Nemo's library for customs officers and visiting dignitaries. We made fold-up chairs for our distinguished crew to use on deck during sunset and sunrise. Other special design specs included a roof over the helm for the helmsman to survive a hundred-foot-high wave like those found

Margaret Augustine, head of the Heraclitus *construction project, takes the helm.*

between Africa and Madagascar. We added a small keel to keep from crunching against coral reefs as well as to help regulate rolling and yawing. The main mast could be used to raise and lower heavy weights for times the *Heraclitus* would be tied or anchored far from any dock. The deck was designed to be strong enough to withstand all the Queen's Ladies of Tonga dancing upon it.

The indefatigable Bill Dempster made detailed hydrodynamic hull calculations with the help of a Boeing specialist, and thoroughly engineered the rig with the aid of Tom Colvin, a master of sail design. Bill designed the GM 271 engine set-up, consulting with the great engine man of the Bay, known to us only as Old Man Parker, whose continuing help was secured by the vivacious Mary Evans. She had said "yes" when asked to take charge of getting the engine installed. No knowledge is much less dangerous than a little knowledge. "I don't know anything about putting engines in ships." "Good," I said. "Find the top man in the Bay Area and convince him that you are willing to learn and carry out whatever he tells you to do." And she did.

Kathelin's directing of the successful European and North African tours of Theater of All Possibilities in 1972 and 1973, helped the first crew on the *Heraclitus* feel at ease on foreign shores, playing for the first time in London, Amsterdam, Berlin, and Rome, as well as in American centers of culture.

An important contributor to the *Heraclitus* was Robert "Rio" Hahn, from the University of Pennsylvania. He has a true explorer's survival instinct and an eye for the significant detail. He was later elected a Director (and then Ombudsman) of the Explorer's Club and is a Fellow of the Royal Geographic Society. A chemist and photographer before stopping at Synergia Ranch on his way to Mexico City, he had apprenticed with the *au point* photographer Minor White and taught a photographic workshop at MIT. Rio and I became lifetime pals dur-

Robert "Rio" Hahn, head of the Amazon ethnobotanical expedition.

ing dark and watchful nights on extended land and sea expeditions in many an entrancing biome, under the dazzling stars of the Southern sky. Rio headed a difficult *Heraclitus* Amazon expedition to study rainforests and the adventure-filled 1983-86 "Round the Tropic World" expedition, which I designed to study sustainable tropic agricultural systems. From these voyages, I produced *Journeys to Other Worlds*, twelve documentaries on special cultures, ranging from Samoa to Ethiopia. Rio did the difficult line production chores for the film, as well as twice rescuing our camera man, once from kidnap by cattle thieves and once from the Soviet KGB in Yemen.

Rio earned his sobriquet from the commander of the Peruvian Navy when he dove to the bottom of the Amazon River (Río), with the captain of the *Heraclitus*, and brought up an important sunken airplane carrying vital intelligence. The Peruvian Navy had been unable to rescue it until Rio arrived with the *Heraclitus* at Iquitos, 2,200 miles inland from the Atlantic. We always had at least one diving specialist on board. Rio managed to bring up the twin-engined Otter aircraft, lost deep below the surface in mud and caught in treacherous whirlpools, by attaching cables to it.

Rio had pulled off several other tricky maneuvers, including rescuing and rebuilding the *Heraclitus* when, in his absence, leaving the narrow harbor entrance of Savaii in Western Samoa, the ship hit a reef during a force six wind. He had to get the ship towed to the one small wooden dock in Savaii, and then learned to weld underwater to fix the gaping hole. Welcomed into the local village, we were given one of their beautiful hardwood floor houses rising two feet above the ground on basaltic foundation stones. Every morning, Rio lined out our day's tasks, and the town crier lined out the work for the villagers. Finally, they offered us a nearby site to make our own village if we would stay. I said, "We don't have enough people to make a village." They kindly offered "a boy for every girl and a girl for every boy." Rio would be the recognized Chief, and I would be the Talking Chief.

ABOVE: Heraclitus *tied off to a great tree on the bank of the Amazon;* BELOW: *some of the* Heraclitus *crew doing ethnobotanical exploration in the Amazonian rainforest. Richard Evans Schultes called on the Institute to take our ship up the Amazon and help preserve the ethnobotanical knowledge of the shamans that was, and continues to be, threatened with extinction.*

These were halcyon days during which we also made a documentary of Samoan life.

The *Heraclitus* is operated by the officers on a volunteer basis and qualified applicants who do a nine-month program to learn able-bodied seamanship, diving, basic theater, and participate in biospheric and ethnospheric research. "Qualified" means ready, willing and able to learn. By the end of the nine months, an applicant should be able to rapidly, perfectly, and easily do what he/she has learned. As of 2008, more than two hundred and fifty people – adventurers, scientists, artists, and lovers of the sea – have completed this difficult program with honor and satisfaction.

Heraclitus *seamanship training covers all aspects of life at sea: learning the ropes, navigation, marine ecology studies, diving, cooking and developing presentation skills through theater and speech. Over two hundred fifty people from fifteen cultures have graduated from this thorough, hands-on nine-month program.*

The *Heraclitus* enabled us to work in the upper Amazon with world-class scientists who later played essential roles in creating the Biosphere 2 rainforest: Richard Evans Schultes of Harvard, who founded contemporary ethnobotany, and Sir Ghillean Prance, who became Director of the Royal Botanic Gardens at Kew, London. Building the ship forged Dempster, Augustine, and me into a top-notch design team. We created a complex ship system, modeled on a synthesis of Synergia Ranch and a traveling theater culture, capable of withstanding sharp impacts and operating in remote regions over long periods of time.

When it came time to build Biosphere 2, the *Heraclitus* played a major role in creating four of the seven biome teams – the rainforest, coral reef, mangrove marsh, and the tropical agricultural biome. To date, the *Heraclitus* has sailed over two hundred and fifty thousand miles, including to Antarctica in 1990, up the Amazon in 1980-81, around the world twice, around South America, the Cape of Good Hope, around the North Pacific, and twice around Melanesia. Several *Heraclitus* expedition leaders are now Fellows of the Explorer's Club or the Royal Geographic Society. The *Heraclitus* has carried the flag of the Explorer's Club on three expeditions.

From 1988 to 1994, the *Heraclitus* engaged in intensive coral reef studies in the Caribbean – especially Mexico, the Bahamas, and Belize – studies that played a central role in the making and managing of the Biosphere 2 coral reef and in biospherian training. An earlier expedition – "Around the Tropic World" – had as part of its mission the study of agricultural systems and subsystems that had survived in the same location for a thousand or more years. From these observations, I could decide which of these subsystems would work best for Biosphere 2's agricultural biome. One of the chosen subsystems was the pig-taro system of Polynesia, and another, the rice-chicken-papaya system from the Mekong River area of Thailand.

The *Heraclitus* also played a major role in the release of John Lilly's dolphins, Joe and Rosie, who had been the chief participants in his human-dolphin experiments. Their luminous, ironic, worldly eyes, educated by seeing so many human antics, really turned me on. John Lilly had allowed Kathelin and me to swim with the dolphins and experiment with Stanislavsky exercises to make emotional center communication. Joe and Rosie performed the exercises, which they willingly accomplished more quickly than we did. So I suggested to Kathelin we try to outdo them by simply hanging there in the water as if motionless. Joe and Rosie circled us two or three times, grokking this strange silence in ordinarily manic humans. Then Joe struck me in exactly the place on the back where the stick would descend when one is faking Zen meditation. Rosie did the same to Kathelin. They were not taken in! I experienced so much rapture from their nimble emotional rapport that my animal totem finally revealed itself. My warrior name, and *nom de plume*, would be Johnny Dolphin.

Captain Claus Tober and Expedition Chief Christine Handte carried the Explorer's Club flag at the start of the "Indian Ocean and South East Asia Expedition" from 1995-2000 under charter with the Planetary Coral Reef Foundation.

East Meets West: the Vajra Hotel CHAPTER FIVE

Oh, East is East, and West is West, and never the twain shall meet …
RUDYARD KIPLING

IN 1976, AFTER THE LAUNCHING OF THE *Heraclitus* and its successful break-in crossing of the Atlantic and an expedition down the Red Sea towards India and Indonesia, I wished to honor my immense debt to the Tibetan civilization that had done so much for my understanding during my 1964 encounters. I also believed that Tibetan wisdom had much to teach the Institute of Ecotechnics. To do this required some complex political, engineering, cultural, and economic organizing.

I initiated a three-party partnership agreement between Synopco, Ed Bass's Fine Line Corporation, and the "Family," a group of Tibetan refugees who had fought their way out of Tibet in 1959. I organized this consortium to develop the Vajra Hotel project, with all its political, legal, cultural, logistical, earthquake, and potable water problems – another big step on the learning curve to building a biosphere. Synergetic Architecture and Biospheric Design Inc. (SARBID), co-owned by Marie Harding, Bill Dempster, Margaret Augustine, Ed Bass, and Robert Hahn, run by Margaret, emerged from Synopco's termination in 1978 to design and construct Vajra. SARBID used only local artisans, materials, and traditions for building the hotel, but consulted world-class engineers for earthquake resistance and purifying water. SARBID later became the prime contractor for building Biosphere 2.

My Tibetan friend Paljor Thondup was held in special regard by several Rinpoches of the different Tibetan schools. He in turn consulted with them on his inner progress. Paljor's Project Tibet in New Mexico helps Tibetans resettle in Western countries. His career had already been followed closely by one of the greatest anthropologists

IMAGE ON LEFT: John sits in contemplation in Durbar Square, Kathmandu, Nepal.

I've known: my friend, the late Christoph von Fürer-Haimendorf. Haimendorf, as versatile in politics and organizing as in the field, served as the advisor on the Gonds to the Nizam of Hyderabad, as head of the Northeast Frontier Agency in World War II, as advisor to the King of Nepal, and was founder and head of the cultural anthropology department at the School of Oriental and African Studies at the University of London. Haimendorf summed up his studies of this remarkable group of Tibetans in his book, *The Renaissance of Tibetan Civilization*, which includes a foreword by the Dalai Lama.

Paljor Thondup in Kathmandu, 1978, IE Mountain Conference. He later launched Project Tibet in the U.S., a non-profit organization dedicated to the preservation of Tibetan culture both inside and outside Tibet.

Paljor and his chief of staff, Dupa Lodou, helped me thoroughly search that uniquely magnificent Kathmandu valley for the best site. Its core turned out to be this Tibetan group's own property near the historic temple, Bijeswori, and the small Vishnumati bridge and burning ghat. All the other hotels were on the opposite side of the river in old Kathmandu. Our site lay directly beneath the gaze of Swayambhu, the yang counterpart to the yin center of the valley, Bodhinath. It was in Bodhinath in 1964 where I had met my first Tibetan master, the Chini Lama, a meeting I describe in my book, *Journeys Around an Extraordinary Planet*.

At that time, the Chini Lama worked ceaselessly at helping the flood of Tibetan refugees escaping from Chinese Communist ethnocide. Having read the lives of Milarepa and Padmasambhava with some thoroughness two years previously, I asked to learn. The Chini Lama pointed out his window to a fourteen-year-old boy leaning against a wall to secure his heavy pack. The boy would be taking messages and presents, in the dangerous monsoon season, to a remote mountain monastery. I was to accompany him. The difficult return, which I had to do alone, nearly cost me my life.

I will never forget standing for half a day in a downpour, sheltered in the hollow of an old tree, eating my last three cold potatoes, and contemplating the long slog ahead down slippery muddy trails. But, what

Vajra Hotel rendering; an East-West-North-South meeting place, a destination.

I learned on that expedition made me feel I owed a great debt indeed to Tibetan civilization. I made an essence vow to repay that debt as soon as I could. The Vajra Hotel in Kathmandu was my repayment, along with helping Paljor start up Project Tibet in New Mexico.

Dupa, whose father had been prime minister of Drongpa in Kham and long associated with Paljor's family, set up a high quality Tibetan rug-weaving studio at the back of the Vajra. Refugees desperately needed worthwhile employment. I asked Margaret to complete my design sketch and build a traditional two-storied courtyard, with a good well in the middle, for Dupa's weaving studio. Spacious high-ceilinged weaving rooms and showroom took up the ground floor, workers lived on the second floor, where the balcony gave them and their children air and light. Sourcing traditional dyes meant Dupa and his artists had to explore the mountains now that they couldn't obtain Tibetan dyes. I showed them how to apply quality control measures. Dupa had an impeccable eye for quality and an under-

Dupa Lodou of Project Tibet.
Versatile partner with Paljor.

Margaret surrounded by Vajra Hotel construction crew: all skilled artisans in wood, marble and brick.

standing of western psychology in selling his carpets. His business grew rapidly and he soon employed more than a hundred people. Later, as Maoism ended, he and Paljor opened up a branch in Lhasa, Tibet. Another member of the group, Thusam, was an expert horse trainer who had gained useful connections with the Nepalese Royal Family when he had broken in one of their much-desired horses.

I envisioned Vajra Hotel becoming a skill-learning center that would enable the Family and related Tibetan groups (numbering about four hundred people), to create a prosperous section of Kathmandu, which I called Vajra Tole. Vajra Tole could be the base for a thriving Tibetan district between Bijeswori and Swayambhu, integrated into the Kathmandu economy. The Tibetans who participated in the Vajra Hotel learned, by working with Margaret, to build their houses nearby in a first class way and to do business in accordance with modern principles.

Margaret synchronized the activities of fifteen master artisan groups, each from a different culture. These groups included descendents of the builders of the Taj Mahal who did our marble work, special furniture makers from Bengal, and the master woodworkers of Kathmandu, who had been slowly marginalized until we showcased their incredible skills. With Phil Hawes as co-architect, Margaret directed the complex project, and, while Bill Dempster worked with the earthquake consultants, Ed Bass worked out complex legalities with a Newari lawyer. TAP put on plays and IE held its Mountain Biomes conference there in 1978 (cosponsored by UNESCO's Man and the Biosphere Program and Kathmandu's Tribhuvan University). Vajra Hotel events quickly became attractions to locals and travelers. Vajra became a destination.

In our usual frugal manner, we took the basement room under the Family's house and set up a small kitchen with a kerosene stove. At times six of us slept there. In a small building next to us was a shower and a hole toilet. Some of the sand and cement mixing crew (a tribe down from the hills) used it as well, and had a merry time scrawling messages on the walls with their excrement. We would cheerfully hose the place down when it came our turn to use the facilities. It is hard to convey the feeling of happiness and good cheer on the building site. Part of it was due to living with our workers, our Tibetan partners, and only going to "downtown Kathmandu," Durbar Square, and the great hotels for special ceremonies and business. Most of it was due to the irrepressible good spirits of the people engaged in doing their craft well. We were building a hotel that was obviously going to add to the uniqueness of Kathmandu.

Margaret managing the Vajra Hotel construction site.

The Vajra Hotel also served us well in our study of tropical forest biomes. I began an association with Tiger Tops, one of the best wilderness conservation camps on the planet. I had met Lute Jerstad, the noted Everest climber, and he introduced me to sustainable wilderness ecotourism. I was flying from Kathmandu to Delhi and noticed how interesting my blond seat companion was; he was concentrating on a Sanskrit text, a rare activity for a Westerner. The

Tiger Tops Lodge, deep in the xerophytic forest, Nepal. Photo by Chuck McDougal.

stewardess drifted by and asked what we wished to drink. The blond guy said whiskey and soda. So, I asked for one too, my first since my union organizing days on the South Side of Chicago. We began talking and became the best of friends. Later, I said, "Lute, I never drink whiskey but I wanted to break the ice with you, you looked so interesting reading that Sanskrit." "I don't read Sanskrit," he replied, "but you looked so interesting that I pretended to read it to break the ice."

Lute introduced me to one of his business partners, Chuck McDougal, one of the creators of Tiger Tops, who later invited me on one of his trips to all the tiger reserves of India (except the Sundabans). Chuck was legendary in the world of big cats; he took solitary walks through the forests, some for as long as two weeks, to study tigers. One of Chuck's tricks was to wait until a tiger slept off a kill and then stalk him to his snoozing place on the back of a stealthy elephant. He took me with him a couple times. Chuck would gaze down at a drowsy tiger and evaluate its condition. I have a copy of one of his photographs – the face of a tiger – in my study. After a few seconds of looking at the photo, you realize that there are no bars between the photographer and the tiger. Chuck had just walked up to him.

Chuck and I were out in a remote area of Tiger Tops when he received an urgent call that a tiger had just eaten a man. On arriving at the spot, just across the river that separated a village from the wilderness, we saw a group of villagers sitting as high up in the branches of a tree as they could. Tigers have been known to bring down a target by jumping as high as twenty feet, some say twenty-two feet. I followed Chuck and his chief guide. On a small trail through the ten-foot-high elephant grass, we found a pile of bones licked clean by the tiger's rough tongue. The only other remnant was a six-inch piece of intestine. A bunch of cut tall grass was strewn beside the bones.

Chuck walked further into the wilderness until he found the point where the tiger's paw prints could be found paralleling the narrow path. The tiger had carefully tracked the man until he neared the open ground by the river, then struck him so hard that his hat had flown upwards to alight on the spear of grass. The man had taken the chance of invading the wilderness because he owed money to the local *bania* (loan shark), and he had been told to get the grass for thatch to help pay off his debts. All the grass outside the wilderness area had already been cut by Indian immigrants who had fled their overcrowded country to settle in the Nepali plains, after the malaria that had kept them out had been eliminated by the efforts of the United Nations. Had the wilderness not been guarded by Gurkhas, all of it would have been gone. Hungry tigers would have been completely replaced by predatory moneylenders.

Chuck caught and caged the tiger for trial, then pleaded its case. Since Nepal had no death penalty for murder, the tiger therefore deserved life imprisonment in a barred and secure jail – the Kathmandu zoo. The authorities trucked the tiger there to serve out his life sentence. The bania received no punishment.

Vajra Hotel, Kathmandu, Nepal.

Chuck spoke at our Institute of Ecotechnics 1979 Rainforest Conference in Penang. While he was sitting in the audience on the first morning, a beautiful scientist, Julia Field, who was also a first-rate tiger and lion tamer, suddenly stopped her talk. She fixed her radiant eyes on the alert man in front of her. Naturally, most of us jumped to the conclusion of attraction, but after a few seconds she said, in a pure naturalist's voice, "You have the face of a Siberian Tiger." Chuck's secret had been discovered. He *was* a tiger.

Vajra Hotel's opening succeeded in realizing my fundamental objective of bringing together in one place representatives of East and West, culturally, and North and South, politically. This synergy of the Vajra Hotel and the ongoing expeditions of the *Heraclitus* gave the Institute of Ecotechnics and its friends and associates remarkable portals into ethnosphere and biosphere. Explorers, anthropologists, ecologists, dancers, the World Wildlife Fund, and other creative individuals and development groups used Vajra Hotel's premises for

Puja by monks from the Sakya lineage in the Pagoda at Vajra Hotel.

encounters, conversations, meetings, and performances. Rinpoches, swamis, depth psychologists, top Asian art experts, Baul singers, European yogis and philosophical teachers reserved its great rooftop pagoda, with its ceiling paintings by Rinchen Norbu and his school of traditional Tibetan muralists.

Norbu was said to be the only painting master in Lhasa who had survived the Red Guard's pillage and violence. Margaret discovered Norbu living in a small room back of the Sakya lamasery, with whose Rinpoche we had close relations. She outfitted him with the means to get his school of painting going again, beginning with contracts to do the Rooftop Pagoda at the Hotel. Like Dupa, he had to find all the minerals and plants required to produce his fantastic colors locally, as it was not possible to for him to visit Tibet in those days.

Connoisseurs of travel considered the Vajra Hotel an architectural gem and "in" place. Knowledgeable German and Austrian travelers picked up on it almost at once, then the French and British, and a couple of years later, the Americans. The Vajra quickly became a sought after destination by travelers to Kathmandu. Many locals, artists, intelligentsia, and business people came for conversations in its dining hall, or climbed up to the rooftop where you could see into Durbar Square, the heart of medieval Kathmandu. Scholars from the West and East sought out our Institute of Ecotechnics library for which I handpicked over a thousand volumes. Its shelves contained the Tibetan canon, approved by the Dalai Lama, and the Hindu canon, approved by my friend, Swami Dharmjyoti, the head of the Nagarjung Order, who became our librarian. I selected an approximation to a Western canon, in literature, anthropology, history, management, and philosophy.

From the Vajra Hotel, you can walk straight up to Swayambhunath (the Self-Realized) Hindu-Buddhist temple complex, rising on a twenty-thousand-year-old artificial platform topping a small mountain. It had been a center of the Naga cosmogony and religion that still exists in Nepal. You're supposed to count the steps as you go up and down, and you have to do it all over again if you lose count. And you can't stare into the eyes of the male sacred monkeys inhabiting the area or they'll charge at you. At the bottom of the stairs are the Buddha's footprints, leading away from the temple toward the open road and the marketplaces of the world. Real enlightenment is a means, not an end.

Exoterically, vajra (dorje in Tibetan) means lightning bolt, but esoterically it means an "indestructible diamond point" of attention where a human can realize *dharmakaya* (the reality or cosmic decision body). If a human being can actually navigate through this ever-changing body using the unchanging nature of emptiness (or no-thingness), the Tibetan masters teach that he/she can communicate with any phenomena in its exact and radiant "isness." I used to go up to Swayambhu every week while the Vajra Hotel was under construction and sit in silence with Sechu Rinpoche, while, telepathically united, we gazed at the Vajra and Durbar Square.

Once, some German tourists staying at Vajra asked to see the fabled Gurkha ceremony where each in solitary turn religiously slaughters a bull with one stroke of his curved, deadly knife. Told where and when to go, they returned complaining that nothing was happening. I could not believe this, so hastened to their square. Sure enough, the ceremony was in full swing. However, the Gurkhas, who stepped forward in intervals, one at a time, would quietly approach the bull as if stalking an enemy sentry, and make such a quick and effective stroke that if you blinked you would never see it. The blood quickly gushed out, and the bull was quickly and quietly removed. There were no blood-crazed crowds, only these mountain men performing their duty to the goddess Kali as part of their metaphysical understanding of the physical universe.

I can never forget the day when some Tibetans rolled into a Kham conference held at the Hotel. These tall guys, swinging loosely from the hips as they floated up our stairs with the ease that comes from being two miles lower in altitude than where they came from, carried their rifles like toys. We sat together sipping yak butter tea until the conversation got friendly enough for them to lay down their rifles. We could break out the barley chang to play chug-a-lug. They had made their way to Vajra to celebrate Paljor's wedding, alert for any ambush that might take them prematurely to their next incarnation.

CHAPTER SIX

West Meets East: Art and Ecotechnics

You can never change things by fighting the existing reality.
To change something build a new model that makes the
existing model obsolete.

R. BUCKMINSTER FULLER

I HAVE ALWAYS THOUGHT THAT METAPHYSICS is essential to understanding human culture, ever since Gustav Mueller from Zurich ("the last Hegelian," as he liked to style himself), introduced me to its delights and dangers in the fall of 1948, at the University of Oklahoma. My practice for over half a century has been to spend at least one evening every week, generally on Thursdays, reviewing some aspect of one or another culture's metaphysical system – a great help in understanding the complexities of the ethnosphere. Since I shared my studies of several of the more comprehensive metaphysical systems, people thought that I was a teaching Sufi, like Idries Shah, or a Gnostic Sufi like Gurdjieff. Some said I was a Tibetan initiate or Taoist yin-yanger, Nietzschean-Alfred Jarry-inspired pataphysicist, Burroughsian visionary, Freudian-Jungian-Reichian depth psychologist, Heisenberg-Einsteinian-Darwinian science mystic, or even a shamanic astral traveler who could enlighten something they considered to be their "minds."

I *have* met teaching masters of metaphysics, shamans and situationists, Sufis and Rinpoches. I *have* made pilgrimages to many, running across others who shared their discoveries. I *have* sat knee-to-knee with a number of physio-psychological masters, one or more from every culture I encountered. I *have* found them or they found me.

I'm sure there are hundreds I know nothing about, but I am always ready to learn. From each of these metaphysical masters, I learned;

IMAGE ON LEFT:
Les Marronniers, Aix-en-Provence, France; site of numerous international gatherings of artists, scientists and explorers at Institute of Ecotechnics conferences.

one master, Idries Shah, a groundbreaking Sufi exponent, insisted that "learning how to learn" was the essential learning. I had two personal conversations with him, one of which lasted a couple of hours and one of which lasted a couple of sentences. They had equal metaphysical impact. Shah's book, *The Sufis*, is a fine example of successful metaphysical transmission. One *pir* took a look at me and hurled a gold and blue club at my head in a noisy *chaikana* in Herat, Afghanistan in 1971. Whatever put my hand up to catch that club before it smashed into me cleared up a point bogging me down at that very moment. The *pir* certainly made several points, among others, that metaphysics, beyond nature, does not cancel out Nature. If you can't catch the clubs, your metaphysics will vanish with your natural disappearance.

Many of the masters I've met didn't graduate from any university of books, but they had all graduated from the university of life. Each master's practical and demonstrated personal power forced me to remember my life and to move nearer to my heart's desire. Madame Jeanne de Salzmann (Gurdjieff's anointed successor) said: "Can you do the three center exercise?" I could and demonstrated it by pointing my finger. "How long can you hold it?" she sniffed.

Some genuine masters do actually run *tekkias* or *dojos* or lamaseries or tribes or useful personal cults or laboratories, but most of the best ones, in my experience, remain anonymous. With "no name," they are left alone to do their thing; often they manifest more qualitative exactitude than is found even in the top professional establishments of the world. Except when one of them happens to be running one, as in the case of Richard Schultes, David Lilienthal, and Konrad Lorenz. These three are masters who operated in the full clamor of life. Superlative qualitative exactitude is the only clue they give. And direct imitation of their action (not activity), open to their direct critique (sometimes only a raised eyebrow or a nuanced gesture), is the only method I've found to access their knowhow. They can't bear to see it done wrong, and if you're open, they'll tell you or demonstrate exactly what you did wrong.

Buckminster Fuller transmitted American transcendental metaphysics better than anyone else I've met with his method of talking to me (and anyone else who listened) at length about everything but metaphysics. Thrangu Rinpoche, Panchen Lama, Shamarpa Rinpoche, T'ai Situ Rinpoche, the Sixteenth Gyalwa Karmapa, Chini Lama, and Sechu Rinpoche each fine-tuned in his own way my attitudes and perceptions with their transmissions of the metaphysics of the school of Swat via the twin transmissions of Tilopa and Padmasambhava. Jazz master Ornette Coleman, writers William Burroughs, Brion Gysin, Lawrence Durrell, and artist Gerald Wilde each showed me how they used multi-dimensional consciousness to transform metabolic and metaphysical functioning.

There are many more masters of mind-body-ego operating in the world than most people realize, though, obviously, few compared to demagogues, rhetoricians, and bureaucrats. They are unrecognizable to those looking for colorful paraphernalia, symbol-filled costumes or striking behavior patterns. It seems to me, a master of metaphysics chooses who he/she wishes to deal with and on what basis. People can't choose a master; thinking they can is an illusion. Once, three Huichol *marakame* showed me how to find an emotional precision that would allow me to survive drastic events. They did this by sitting in the cloud of dust raised by the dancers' feet. At first, I couldn't figure out how they breathed. Then I noticed, by imitating their state of consciousness, that when they could see small pockets of clean air floating by, they would take a deep breath in from that patch of air.

All this talk of real masters is by way of introducing the next phase on the road to Biosphere 2 and the role of the Institute of Ecotechnics (IE) in its realization. We had started our series of confer-

Mark Nelson speaking about the Institute of Ecotechnics Conceptual Model at the Planet Earth conference, 1980.

ences exploring biospherics with a small gathering in the dome at Synergia Ranch. For our first total systems approach, we began locally with the "Ecology of the Upper Rio Grande" conference. We formulated our task: to combine the pleasurable with the profitable, the beneficial with the beautiful. We established principles. We invited only outstanding scientists, thinkers, and artists with widely recognized independent accomplishment. We did not invite the media, whose pressure for one-liners might diminish the speakers' thoughts to sound bites and "lead" sentences.

We wasted no time hammering out manifestos that all participants agreed to. Our motto: *What's in it for you is what's in it.* Each individual would leave with whatever he got out of it. IE members prepared *haute cuisine* from food grown on the ranch, arranged the conference ambience, and handled all logistics; this raised the quality and lowered the costs. Spontaneous group and one-on-one conversations were a primary working tool in between the talks and questions.

Each speaker was given plenty of time to make a presentation and field questions.

Our second conference took up basic problems in Communication. We invited William Burroughs. His cosmopolitan background gave him the depth of resources to overcome the difficulties of using words only as stimulus-response conditioning devices. William had developed cut-up montage and riffing techniques to liberate unconscious subtexts from the cant of *Time*, *Life*, and *Fortune* magazines. We also invited Hank Truby, an expert in communicating with dolphins, and inventor of a device that allows one to identify a speaker from his voice patterns. Hank was well known for his detailed study of French village *patois*, demonstrating the importance of sublanguages in creating local cultures.

We sat around the tile-topped pine tables under the portal of Synergia Ranch's dining hall in that hot autumn and talked animatedly between sessions. William had shown up with his learned, conscientious, and hip literary executor, James Grauerholtz, with whom Kathelin and I made a lasting friendship.

William became more than a literary hero; he became an essence friend. He participated in a particular part of the Biosphere 2 design, as he was something of a specialist on lemurs and galagos. We considered his advice in our final selection of galagos as our representative primate. I and Kathelin adapted a play, *Deconstruction of the Countdown*, based on William's and Brion Gysin's work. He and Brion came to see the Theater of All Possibilities' production of it at the American Center in Paris.

Later, Kathelin and I visited Burroughs in Lawrence, Kansas where he lived with his cats. In the spacious vacant lot back of the house, we meditated on the "Third Mind" created when two share a vision as he and Gysin had, and on the "Process" needed to deal with the

TOP: *William S. Burroughs made us hear the thunder of the "Four Horsemen of the Apocalypse" at the 1980 IE Planet Earth conference.*

BELOW: *Dr. Paul Rotmil, who helped eradicate smallpox, and Burroughs converse over lunch at the meeting.*

"Word Virus." William, dressed in his army combat jacket, casually tossed up a sheet of paper, then deftly sliced it in two with his razor sharp plastic knife as it fell back toward our three cups of tea that he had prepared with aristocratic precision and spare elegance. He showed us his Bowie knives, an oiled revolver and a shotgun. He unloaded and re-loaded his thirty-eight with care. "I never leave a gun unloaded," he said.

I had a flashback to a scene in his white painted Bunker on the Bowery. On my return from a trip to the Amazon, Burroughs had pulled out his collection of blow guns that delivered darts poisoned with curare and said, "Wanna shoot some?" We blew dart after dart into his great wooden door. Burroughs believed that only the most disciplined, daring, and imaginative alliance of humans could stop what he called "The Nova Mob" from blowing up the planet and moving on to find new victims.

After the first two conferences succeeded, the Institute decided to convene ten world class annual conferences, taking as subject matter a series of key biomes and their interpenetrating surroundings — desert, ocean, grasslands, rainforest and mangrove marshes — and then conclude with a complex interweaving of small area biomes in high mountain ranges. Small biomes (because of the quick changes in climate due to changing altitude) were the closest natural analogue to my idea on how to build a complex biosphere.

For the next five years, we met, listened, and talked with outstanding artists, top thinkers, and scientists. After these most enlightening and enlivening conferences on Earth's biomes (one of them was published as *Man, Earth, and the Challenges*), the IE series continued on to: Earth (Geosphere), the Solar System, the Galaxy, and the Cosmos, in the spirit of Vernadsky's dictum that, "The biosphere is a geological force and a cosmic phenomenon." The tenth was on the biosphere itself. The biospheric system cannot be understood without

Institute of Ecotechnics conference on Planet Earth, 1980, Aix-en-Provence, France.

considering its total context of input and output of materials, energies and behaviors. As the famed physicist Enrico Fermi once demonstrated (with sweeping motions of chalk on a blackboard) to a few of us at the Colorado School of Mines, particles can be whipped around the galaxy and thrown at high velocities through the biosphere, perchance to cause a fateful mutation.

IE held its first world scale canonical conference at Synergia Ranch in 1976. We were hosting experts like Mickey Glantz from NCAR, Boulder, and Michael Evenari from the Hebrew University, Jerusalem, who presented "Deserts in the Biosphere." To our dismay, we had scheduled our conference at the same time the United Nations had scheduled their prestigious world conference on deserts at Nairobi, Kenya. Since all our invitees were at the top of their fields, we were gloomy. Well, we thought, we could always read the published papers of the invited speakers while they spoke in Nairobi. Incredibly, they all showed up. Mark Nelson and I asked them why they came to our affair instead of to the top meeting in the world.

They laughed and looked at one another. One said, "Well, we knew nothing would happen there. We would hear the same things as always. Here … there is a chance something new might occur."

Michael Evenari joked, "You said I would get to see the Chihuahuan Desert. I do not think this is a desert. By the look of your vegetation, you have only within two centimeters, on average, less rainfall than your evapo-transpiration. Yes, technically it's a desert, but you must come to my Negev to see real desert." Mark Nelson took Michael up on his offer and participated in Evenari's great work restoring the destroyed rain-water catchment systems of the Nabatean civilization. And Robert Hahn made a pilgrimage to the Negev to observe scorpions with that keen eyed master.

Later, after 1980, IE convened nearly all its conferences at its European experimental farm project in the Mediterranean biome at Les Marronniers, six kilometers north of Aix-en-Provence in France. We contributed design, labor, and consulting to this fifteen-acre working organic farm and conference facility in 1976. We had the help of the *Heraclitus* crew who had just docked the ship at the Vieux Port in Marseilles after going through the Panama Canal, up the Caribbean and across the Atlantic.

The Les Marronniers property had vegetable gardens, field crops, orchards, grazing, woodlands, and two artisan workshops surrounding a two-story house. Its grounds were laid out in the classic style of the last years of Louis XIV. An annex provided a genial space for theater, acting classes, and Ecotechnics conference sessions. It was the former atelier of Monsieur Jouve, the master potter who instructed Picasso when he decided to paint on plates. The atelier of Cézanne was a few kilometers away, and our water came from Mont Saint Victoire. Plantane trees, chestnuts, pines, bamboos, firs, and one lone cypress gave shade from the hot sun and protection from the fierce Mistral wind. Blackberries, figs, grapes, chickens, sheep, and the

Les Marronniers, organic farm and conference center near Aix-en-Provence, France.

potagerie gave an air of plenty. We planted poplars to restore the old woodlot.

Margaret Augustine elegantly manifested my concept for a two story *Place des Voyageurs* to be built over the old hog house to enable Les Marronniers to accommodate fifty-five to sixty people. We planted a cherry tree in its center. Participants would be composed of IE members, speakers, and interested public. Twelve to fourteen speakers is our usual number for three-day conferences. *Le Place* allows Les Marronniers to meet its expenses by being open during the *belle saison* for use by theater, artistic and other innovative groups that utilize the facility for their workshops.

Margaret's vivacious sister, Molly Augustine, has run Les Marronniers since 1981, with her co-manager and husband, the well-known sculptor and painter, Cesco Rimondi. He masterfully blends his two heritages, Ethiopia and Italy, painting and sculpture, with ecotechnics. Cesco also runs Les Marronniers' wheat fields and gardens, increasing their fertility by judicious recycling and the use of legumes.

Molly Augustine and Cesco Rimondi, directors of Les Marronniers.

The vegetable garden provides fresh produce for the conference center.

The Mediterranean biome is only found in three other areas of Planet Earth: at the tip of South Africa, in Southern California, and in Southwest Australia. Although far from the greatest in extent of the biosphere's biomes, the ecology of the Mediterranean biome holds special interest. Its bioregions in France range from perfumed garrigues to rugged maquis, from deep limestone caverns to the slopes of snowy peaks. In the Mediterranean, Provence was "the Province" to the Romans, who thought it stood out from all others. The whole world seems to love Mediterranean biomes, and their citizens invade, trade, vacation, or pilgrimage there.

Cultures thrive and reach remarkable climaxes in the Mediterranean biome. I have found pure delight dancing with dervishes in Morocco, sipping Egyptian coffee in El Fashawi in Cairo after contemplating the Mamelukean Mosques, browsing the Street of Perfumers, passing through the transdimensional wall at the Sybil's power place in Cumae (the same, it's said, that Odysseus had gone through to visit

Tiresias), savoring the inner sea's goodness concentrated in Marseilles Vieux Port bouillabaisse, gazing at the wave foam on which Cyprian Aphrodite had glided to that island's shore, and jolting in the wine dark sea's steep-waved power while sailing the *Heraclitus* from Toulon to Bonifaccio. I have loved every bioregion's special perfumes that render those shores so fragrant that even an adventure we nearly pushed too far remains redolent in my memory. Five of Biosphere 2's leading *dramatis personae* – Marie Harding, Kathelin Hoffman, Ed Bass, Bill Dempster and I – hiked through the scorching, rugged, Corsican back-country from Bonifaccio to Ajaccio. Along this route, I imagined guerrilla tactics used by Paoli and Bonaparte, but we saw only blackberry-covered ruins of once flourishing villages where we found no running springs. Arriving thirsty and exhausted in the darkness in Ajaccio, we dug foxholes in the beach sand to avoid the wind and slept soundly through the night.

The Institute of Ecotechnics first conference at Les Marronniers was "The Ocean," timed to coincide with the arrival of the *Heraclitus*, which we had launched in San Francisco Bay the previous year. The ship had a stop scheduled in Marseilles en route to studies in the Red Sea and Southeast Asia. The *Heraclitus* crew used their time in France to play a big role restoring the neglected *maison* for the conference. Synergy in action: *Higher, Gayer, Simpler, Easier*. The goals Stanislavsky proclaimed every actor should enshrine in his heart.

In 1978, we started the process of registering the Institute of Eco-technics in the United Kingdom, the better to work internationally. The UK is not only a center of synergy, the Commonwealth culture of over forty nations, but it was becoming part of the New Europe culture, as well as maintaining its special relationship with the USA. Three great universities are within easy reach of London. Libraries and museums offer tremendous intellectual resources. London makes a perfect location for planetary wide activities.

We looked at Paris, New York, or London to locate the Institute, our center for science, and October Gallery, our center for art. I came to the conclusion that in Paris you needed to be part of a clique to support efforts in these fields and this would automatically arouse opposition. In New York, a new approach could sink without a trace or make a big hit, but even if it was a hit, you could soon disappear unless you were always "doing something new." In London, I met an intelligentsia that had interest in ideas, forms, and history. London would take longer to break into, but it seemed that after "the Brits" fully checked you out, you were accepted for what you did, to a greater extent than the other two great cities.

Not that I wouldn't still want to hang out at the Café de Flore or Les Deux Magots and visit with my French friends for their *je ne sais pas quoi*, their civilized eye for style and authenticity, their wit and *savoir faire*, and wonderful dinner conversations.

Not that I wouldn't still visit New York's Manhattan and look forward to encounters in the street, and experience that energy, that edginess, that willingness to experiment with life, and contact those wounded hearts who knew no limits to their hopes, their wild flights of eloquence or their desperation.

In Paris everyone who could had a worldview. In Manhattan, everyone who could viewed themselves as a world. In London ... well, in London everyone had their local hangout (from pub to club) and a park to stroll in, and could find someone, somewhere, who knew all about anything they needed to know or how to get there and back. In Manhattan, you could become part of the scene, in Paris, part of civilization, in London, a part of London. Anyhow, that's how I saw it and still see it.

It was in 1977 I conceptually designed the October Gallery, co-founded with Marie Harding, Kathelin Hoffman, and Margaret

Lundonia House, nineteenth century Victorian schoolhouse, now home of the October Gallery Trust, Bloomsbury, London.

Augustine, a "transvangarde" art and events center that opened in a three-story 1863 classic brick building originally solidly architected as a home for foundlings and lost children. The gallery is just a minute's walk from Queen Square where T.S. Eliot presided over Faber and Faber's famous poetry division. It was where Queen Caroline took food and read to George III who had been confined in the Hospital. Just up from Faber and Faber is one of the world's centers of brain research, and across from there is the world-renowned Royal Homeopathic Hospital.

A few blocks away, at Gray's Inn, begins the legal district of London, and down an alley street across Southampton Row lies Russell Square and the University of London with its School of Oriental and African Studies (SOAS). We have had much interchange with SOAS, and especially with the founding head of its department of cultural anthropology, Christoph von Fürer-Haimendorf, who combined Vienna, Oxford, Kathmandu and Hyderabad in the most remarkable fashion.

October Gallery opening, 2005 Transvangarde show. Poet & photographer Ira Cohen (bottom left) with camera. Other artists in the show, photographer Carol Beckwith, and digital artist Steina Vasulka, appear in the crowd.

Gerald Wilde, 1955. UK's great abstract expressionist. Photo by Gilbert Cousland.

It took several months of staying in London to find a place we could afford. We needed a space large enough to do our world city theater and art gallery project that was also a place where a person from any class or culture would feel inspired to visit if we put on something of interest. We wanted a place where we could feel at home. Such criteria made the Bloomsbury neighborhood, a traditional center for the intelligentsia (particularly writers), our first choice.

We opened in 1979, with a major show of the work of artist Gerald Wilde. I had met Gerald when a wonderful philosopher and historian friend of mine, John G. Bennett, invited me and TAP to his new school at Sherborne House, in Gloucestershire. We arrived at Sherborne after a visit to Dublin to see the Book of Kells and to visit the sites made famous in James Joyce's books.

The October Gallery was set up as a charitable trust, dedicated to the advancement and appreciation of art from around the planet and to the exhibition and promotion of the transvangarde. The transvangarde can be likened to antennae of the ethnosphere, gathering and disseminating vital information about ongoing actions as and even before they occur. Transvangarde artists come to the October Gallery from dozens of countries and have helped the Institute of Ecotechnics make expeditions to some of the most intriguing but difficult-to-travel-in bioregions.

Chili Hawes moved from Santa Fe to direct the October Gallery. She continues with her team to provide the guiding spirit for the complex program of artist showings, readings, theatrical events, seminars and music, and has served as secretary and currently treasurer of the Institute of Ecotechnics.

Deborah Parrish Snyder, a new key player in ecotechnics, entered into the action at that time. She completed the ecotechnics program at Les Marronniers in organic gardening in 1981, and in 1982 worked on our savannah restoration project in the Kimberley, West Australia, at Birdwood Downs station. While in London in 1983, Deborah began managing Synergetic Press, which had been started by Kathelin Hoffman back in 1969. Deborah expanded the publishing company and continues to run it to this day. She also became a director of the Institute, helping to manage the annual Ecotechnics conferences.

While at Birdwood Downs, Deborah operated the ham radio for the twice daily call to our other station, Quanbun Downs. In those days, the radio was the only form of communication for the outlying stations. Birdwood's radio answerback was "Tango Foxtrot Tango." The Aussie lads from the other stations listening in liked her warm voice and said that they were going to call her by our code name. She said "You can call me Tango, but you can't call me Foxtrot!" Tango, as she became affectionately called, could drive a tractor as gracefully as she could dance.

Chili Hawes (left), director, and Elisabeth Lalouschek, artistic director of the October Gallery.

Deborah Parrish Snyder began managing Synergetic Press in London, 1984.

Birdwood Downs paddock, cleared of invasive scrub trees and replanted in improved grasses and legumes.

Australia's Outback attracted me with its last chance for me to emulate my ancestors and take part in large scale frontier type ranching. Biospherically speaking, cattle raised on a savannah too climatically harsh for agriculture and with competing natural big grazers (as in East Africa), if overgrazing is avoided, are an addition to the complexity of the food chain. Their meat is deliciously and healthily different in composition from feedlot cattle and of course they enjoy their life moving around different paddocks (pastures).

In 1975, an opening came to obtain a station in the Kimberley of Western Australia for three of my colleagues to make a partnership on, which I highly encouraged. We ran about seventy horses and three thousand head of cattle and muster (round-up) consisted in its most glorious part, of herding a thousand head across the fabled Fitzroy River (in the dry season of course). It was up at four in the morning and turning in with your swag under the Southern Cross and a glittering galaxy (like when I was a boy in Oklahoma before the pollution came and fuzzied the night skies).

Cattle muster (round-up) at Quanbun Downs station, near Fitzroy Crossing in the Kimberley.

Later, on my way out from Aussieland, I met a guy at Derby's Boab tavern, and made a $20 dollar option payment for a five-thousand acre parcel we later called Birdwood Downs, because we raised Birdwood Grass (sent by Field Marshall Birdwood from India to help re-grass tropical Australia). Mark Nelson and I had worked out a method to regenerate its former grassland devastated by overgrazing and now overrun by invading species. This station is now managed by Robyn Tredwell, a native Australian and my versatile friend since she dropped in to help out at the Vajra Hotel, and then became ethno-botanist collector for the *Heraclitus* Amazon expedition. One time, our row boat was caught in an Amazonian whirlpool and round and round we went until we cautiously extricated ourselves.

Robyn was born and raised in rural Queensland. Formally trained as a nurse, she did bush nursing in North Queensland, around Cairns and Cape York Peninsula. In 1978, she went to Nepal and financed and set up a health clinic for the Bijeswari Tibetan refugee community in Kathmandu which is where I met her. In 1980, Robyn joined

Robyn Tredwell in the Amazon,
1982.

the Institute of Ecotechnics expedition on the *Heraclitus* to the Amazon River, where she managed its ethnobotanical collections. She continued to voyage on the ship until 1985 visiting over twelve tropical tribal cultures, continuing with her collections of plants of medicinal and agricultural interest in Central America, Pacific Islands, and Indonesia. In 1986, Robyn arrived at Birdwood Downs where she quickly took the reins as the project director. With Diana Mathewson's leadership, we developed a new horse breed better suited to work in the tropics, a mix of Arab and Quarter Horse. A couple of other Australians had the same notion and now the Quarab is a recognized breed. It takes a special kind of gaiety, resilience, stoicism, ingenuity, and never say die to live in the Outback with its three seasons of Flood, (Hot) Drought, and Fire. Robyn has all of that, and my few weeks a year there always regenerate my spirits and health. In 1995, Robyn received the Australian Rural Woman of the Year award in recognition of her work in sustainable landcare.

Birdwood Downs is perhaps one of the craziest projects I've been involved with, but is the closest thing to the vibration of my grandfather's ranch. Since 1978 when we started work there, over one thousand acres of pasture have been cleared and planted in improved grasses. Today it is home to the Kimberley School of Horsemanship offering courses in all aspects of horse riding and care. I wrote this poem about the project; one of many inspired by the place.

SAVANNAH SYSTEMS, WEST KIMBERLEY

Bougainvilleas blazed red back at heavy sun.
Cockatoos cascaded a wary way down into the
  big bloodwood branches.
Bauhinias offered shade and food to Brahma upfeeders.

Gessie Houghton and John after a morning of wattle clearing in Victory paddock.

Peacocks began to spread stiff jeweled tails for
    aloof peahens.
Later, horse hoofs clomped at sunset wondering
    what to do.
While crosslegged on bales of mulch hay beneath
    archaic Boabs
Popping our champagne while day disappeared,
We loafed, intelligently wafting on happiness.
Under sleek cool stars we hiked back over wattle-
    cleared grass
To banana bread, coffee, and laughter on the veranda,
Waiting for fresh tender brisket with Robyn's frozen
    mango dessert
To savor immense happenings during daily routines.

Outback Australia and East Africa gave me the experience and knowledge to model the savannah for Biosphere 2, a model that Peter Warshall and Linda Leigh realized in marvelous detail.

For over two decades, the Institute initiated and helped set up a series of biomic projects around the planet. We also held a series of international conferences each year. These first focused on the Earth's major biomes and larger orders of organization (planets, solar system, galaxy, cosmos) and then a series concentrating on the new science of biospherics and its major tool, closed ecological systems. Several of these conferences were held in locations other than France: The Vajra Hotel in Kathmandu, Nepal for Mountains, co-sponsored by UNESCO's Man and the Biosphere Program and Tribhuvan University (1978); in Penang, Malaysia, for Rainforests, co-sponsored by the Man and the Biosphere Program and the University of Penang (1979); in Perth, Australia, on Transition Zones (Grasslands), (1980); one in London in 1987, one in Krasnoyansk, Russia in 1989; in 1992 at Biosphere 2 in Oracle, Arizona

Institute of Ecotechnics Mountain Conference, Kathmandu, 1978. Lute Jerstad, famed climber of Everest, speaking on "Mountains as Adventure."

for Biospherics; and at the Linnean Society (part of the Closed Ecological Systems Biospherics Conferences) in London in 1996.

IE acquired first-hand knowledge from the highly qualified men and women who participated in the conferences. Some became quite interested in our ideas and approaches. Feedback from these authorities inspired us to improve our work. They demonstrated standards of thought, procedures, and attitudes that the Institute endeavored to emulate. We invited far-seeing thinkers and explorers, such as Moon, Planet, and Asteroid experts James Head, Mike Duke, Wendell Mendell, Eugene Shoemaker, and astronauts Joe Allen and Rusty Schweickhart, Astronomers Bernard Carr and David Malin, Kraft Ehricke (co-adjutor to Werner von Braun), Alexander King, co-founder of the Club of Rome and Secretary of the OECD, my old friend Bucky Fuller, and versatile planetary explorer Nigel Winser of the Royal Geographic Society. Russian and Indian scientists began to contribute to the knowledge base.

Workshop leaders in first Biosphere 2 design conference, Sunspace Ranch, Arizona, 1984. TOP LEFT: *Jack McCauley, Loren Acker, Rusty Schweickart, George Mignon, David Wright, Ralph Abraham, Carl Hodges, Keith Runcorn, Al Hibbs, John Stolz, Clair Folsome.* BOTTOM FROM LEFT: *Carol Breed, Peter Warshall, Don Paglia, Roy Walford, Cameron Gunderson, Mark Nelson, Joe Hanson, Tony Burgess.*

In 1984, the climax of ten years of enlightening conferences was the conference about Biosphere 2 at Space Biosphere Venture's new Sunspace Ranch—the only IE conference to be specifically action-oriented. We convened to answer the one basic question my mentor, David Lilienthal, insisted be answered: *Is it do-able?*

The three dozen top scientists and engineers who attended that 1984 conference in Arizona agreed, that yes, Biosphere 2 was do-able and, more importantly, they agreed to get on board.

We convened the first international conference on closed ecological systems in 1987 to gain the support of experienced Russians like Oleg Gazenko, the indomitable Evgenii Shepelev, and cool Josef Gitelson, of the knowledgeable Odums, Eugene and Howard, and of the sophisticated Keith Runcorn who made it possible to hold the conference at the Royal Society in London, one of the few places Soviet scientists could get permission to travel to. Keith brought us his poised associate, William Chaloner, head of the Geosphere-Biosphere committee of Great Britain. We needed their rigorous scrutiny of our research program, the data from the Biosphere 2 Test Module experiments, and my overall design. Without passing their keen, experienced assessments, we would have to go back to the drawing board. I mean *computer*, for during that ten-year period, "drawing boards" had vanished to the museums of antiquities.

We had done it. The conferences not only served as a first-class tool for the acquisition of knowledge, but also enabled us to form a top scientific cadre around the mission of creating a unique scientific apparatus—Biosphere 2. Many of the scientists became close friends and consultants and all sharpened our understanding. Besides the eight mentioned above, I must acknowledge ethnopharmacobotanist Richard Evans Schultes, ethnobotanist Sir Ghillean Prance, astronaut Joe Allen, gerontologist and pathologist Roy Walford, astrophysicist Al Hibbs, geologists Edwin McKee and Jack McCauley,

John Allen tours with Evgenii Shepelev (center), Oleg Gazenko (fourth from left), Josef Gitelson (far right). Biospherians Linda Leigh, Mark Nelson (behind), Deborah Parrish Snyder, director of publications (far left).

explorer Thor Heyerdahl, ethologist and ethnologist Bernd Lötsch, conservationist and environmental chemist Balkrishna Tejam, the great microbiologists Lynn Margulis and Clair Folsome, the planetary thinker, James Lovelock, and virtuoso of chaos mathematics, Ralph Abraham. Many other wise and incisive minds contributed fuel to the fire.

ON LEFT: *Richard Evans Schultes; and* RIGHT: *Keith Runcorn (center) and consultant, Stanley Buchthal (right), with John on tour of Biosphere 2.*

Ventures in Timber and Jazz

CHAPTER SEVEN

The dogs may bark, but the Caravan moves on.
SUFI PROVERB

AFTER FORMING SARBID (Synergetic Architecture and Biospheric Design Corporation), I needed to make one more organizational step before building a biosphere. A special corporation was needed to supervise the financing and engineering. I initiated and co-founded Decisions Team, Inc. (DT), with five investors – Margaret Augustine, Marie Harding, Mark Nelson, Kathelin Hoffman and Ed Bass – each of whom owned twenty percent. For a while, Rio Hahn was also a director and part owner. On principle, I never owned shares in any corporation I founded since closing down my Mountain and Manhattan coal enterprise. The owners of DT purchased enough stock to get the company up and running, and appointed me its strategic director, Margaret the CEO, and Ed the chairman. Marie would be chief financial officer.

John during construction of Caravan of Dreams, Fort Worth, Texas, 1981.

The DT board would arrange the financing, and direct any projects whose ideas had passed the Institute of Ecotechnics critique sessions. The work also had to be judged by the Theater of All Possibilities. I figured that these three companies, open to TAP's feedback, would, with a few more years experience, be capable of handling any project, including working with the National Aeronautics and Space Agency (NASA) as a subcontractor on artificial biospheres. In 1978, it seemed certain that NASA would, before too long, build a prototype biosphere for Mars exploration. How wrong we were!

Up until now, I had financed my technical and artistic ventures with the sale of stock, loans secured against property, and donations of intellectual capital and skilled labor. I designed projects as learning

IMAGE ON LEFT: *Las Casas de la Selva (Houses of the Forest), Puerto Rico. Director, Thrity Vakil, inspects line of mahogany trees planted in reforestation of highly eroded coffee land.*

situations where, in principle, everyone did everything. I always required my investors to be interested in and capable of contributing some work in their specialties. I didn't want absentee owners. I joked with them that they must, besides giving their do-re-mi, also put in fa-so-la. Kathelin joked back *si-si-si*. Our *esprit* attracted many creative people without whose participation the projects could not have been realized. I called the company Decisions Team to indicate that it would not be driven by political or market forces, but by the need to supply something that was not yet in demand, because no one had thought of it. We would have to supply those needed "somethings" with a quality that would create the demand.

However, shortly after a presentation that resulted in the formation of DT, Ed Bass said he wished to discuss the results of our earlier efforts. Ed shared the ownership of the Vajra Hotel project with Marie Harding (who represented what became the Biospheric Design people), and some of my Tibetan friends, who contributed the property and local support. Margaret and I donated our management and design time. In 1978, as I was seeking investors, Ed told me that he was forming a venture capital corporation, Decisions Investment (DI), which might be interested in certain Decisions Team projects.

DI offered a deal to DT for the right of first refusal to participate in financing any of our proposed projects. If DI chose to invest and DT agreed, Ed would direct the finance committee and be in charge of the budget. DT would manage the projects, heading a consortium with Biospheric Design and the Institute of Ecotechnics. On each project, there would be space for TAP to make its contributions towards understanding the inner dynamics of the project.

If Decisions Investment decided not to join a project, then Decisions Team could arrange to access capital elsewhere. In case disagreements prevented a joint venture from going forward, either DI or DT could buy out the project from the other, or they could put it up for sale and split the proceeds. DT's board (chaired by Ed Bass) would rule on all questions brought before it by consensus. DT's track record for reaching consensus on widely different projects convinced me that this method was workable.

Ed said he could put forward a small percentage of his assets in long-term, ecologically-oriented projects because of the returns that might eventually accrue from such projects and their spin-offs. His business affairs might profit from investment decisions made with up-to-the-minute, real knowledge of the planet, and from meeting with the world-recognized experts who worked with us. This privileged knowledge could alert him to further business opportunities and would also broaden his choice as to which non-profit ecological or-ganizations to donate to.

When Ed first met us in 1973, Richard Rainwater, the Bass Brothers' financial genius, stated that they were worth about forty million dol-lars. In 1978, when Ed talked about doing a joint venture with DI in which he didn't wish to invest more than a "small percent" of his as-sets, Marie and I thought Ed was talking about investments of at most a million dollars. While serious money, a million dollars would only allow the smallest start-up for real work on a research biosphere. I therefore needed to interest a number of individuals to invest in my proposed Biosphere 2 Test Module and to set up a research center.

I also kept looking for relatively small-scale projects that would help us gain practical knowledge about the specific biomes that would be part of Biosphere 2. I required more knowledge to make a detailed design for Biosphere 2, especially the rainforest, which is the most complex of all. I thought it was the most important biome next to the agricultural one. The first of our joint venture projects was what I called Las Casas de la Selva (Houses of the Forest), a thousand-acre rainforest enrichment project located in Puerto Rico.

Las Casas de la Selva rainforest enrichment, sustainable forestry project.

I located the site with the help of Puerto Rico's Department of Development. After we had explored the whole island for possibilities, we found this site abandoned by the local *jibaros* (forest mountain farmers) because of its steep slopes, heavy erosion, and lack of roads. Mules took the jibaros with their bananas, coffee, and logs in and out on steep and narrow mountain trails. It needed about $400,000 for the land purchase and set-up: the planting of forty thousand hardwood trees, fitting them into the existing forest, and the building of staff quarters, a library, woodwork shop, and sawmill. DT's contributions could reasonably be valued at a half-million dollars.

The Department of Development, which liked and supported my IE-approved plan, considered that we could help Puerto Rico get out of its trade imbalance through exports of timber. I had looked at numerous rainforests to find the best place for us to start this project, including expeditions to the Amazon, Chiapas, Mexico, and several Indonesian islands. Robert Hahn had even traversed the African

Congo. But Puerto Rico had real advantages for our type of biome research-oriented project: there was no malaria, no yellow fever, no poisonous snakes, and there were American property laws to protect investors. Another big plus was that a number of vibrant Puerto Ricans, including the master sculptor and gardener, Santtos Torres, grokked what we wanted to do and gave us wonderful assistance.

The U.S. Forest Service and Puerto Rican Forest Department became interested in our project, supplying the tree stock (mahogany and mahoe), and assisting with the planting costs. Our property was beside their Carite Forest Reserve and so would help protect it. We found and planted some endemic hardwoods, species threatened from earlier over-exploitation of the island's resources. Today, we keep about five hundred acres of the steep-sloped area as a wilderness, occasionally visited on foot by ecotourists. My approach to growing hardwood timber is to plant at twenty-foot intervals, as is conventional, but without clear-cutting the forest. Fourteen-foot-wide strips are left completely undisturbed, while six-foot-wide planting strips are left relatively undisturbed so the forest mychorrizae and bacterial life are left intact. The trees then grow straight up, seeking the light. Only a skilled ecologist could tell that he was on a plantation, as opposed to a natural rainforest.

Line planting of mahogany trees in otherwise undisturbed forest.

Las Casas would play an important role in stocking Biosphere 2's rainforest (particularly its cecropias, heliconia, and coqui frogs), as well as in training the biospherians in rainforest ecology. Working on this project for a couple of months a year, along with the *Heraclitus* expedition up the Amazon, allowed me to figure out exactly how the Biosphere 2 rainforest should be laid out. Now recognized as an auxiliary national forest, Las Casas has had capable assistance in its data gathering from Earthwatch Institutes's dedicated volunteers. Tropic Ventures, the corporation that organized Las Casas, was the first joint venture of DT and DI. The experiment assured the directors that this consortium could operate effectively.

With Las Casas de la Selva up and running, marking the completion of six years of IE-targeted biomic programs, Ed Bass surprised us in 1980 by initiating a much larger investment from DI to DT – several million dollars. He wished the board to consider Fort Worth as a site for a second joint venture. IE and BD definitely had no such project in mind at that time, being focused on biomes and biospherics. We had covered the city biome by starting up the October Gallery.

Still, for years Kathelin Hoffman and I had nurtured the concept of an avant-garde arts center that combined theater with a jazz and blues club. We had explored Berlin and Chicago as probably the best sites. Fort Worth was certainly not the first city that came to mind. Encouraged by his brother, Sid, Ed had bought several dilapidated properties in the downtown area, where the Bass Brothers already owned quite a few. Ed told us how he wished to develop a new property his brother had suggested Ed buy, using some of the recent profits they had made through Sid's huge deal with Texaco.

Ed figured as part of his long-term investment strategy, the Theater of All Possibilities could carry out its Caravan of Dreams concept and help rehabilitate Fort Worth's downtown. The lively new venue would also enhance property values, since it would bring a vigorous nightlife to the area, attracted by big name artists like Ornette Coleman, Wynton Marsalis, Eartha Kitt, William Burroughs, John Lee Hooker, and others. It would be a great place for Ed to step out, on occasion, and he said it would pay back his debt to his hometown.

The leitmotiv of our Vajra Hotel project had been to create a world that would serve as a gateway into Mountain Asia, The Rooftop of the World – a gateway that people of all types and cultures could enter from all directions and go out again with a bit more magic in their glances; a gateway like the Afghanistan of old, caravans but without the camels.

Caravan of Dreams jazz & blues nightclub, theater, rooftop bar, downtown Fort Worth, Texas, 1983.

Our leitmotiv for the Caravan of Dreams was to create a world in which avant-garde and mainstream America could meet face to face, to design a space worthy of the accomplishments of great performers and connoisseurs of jazz, blues, and freeform. Kathelin and I called it the "Caravan of Dreams," a name that came from a play I had written based on stories from the *Arabian Nights*. Marouf, the hero, had only the legend of his Caravan of Dreams to inspire him until one day the caravan actually showed up, as it crossed the desert from the Unknown Country.

Our Caravan of Dreams would be a place where William Burroughs would feel able to speak, Ornette Coleman to compose, Howling Wolf and the Chicago Art Ensemble to play, Molissa Fenley to dance, the Theater of All Possibilities to perform, Texan Metroplexans to enjoy the waterfall in the rooftop bar, and I could write in public on a table just like at the Café de Flore. Kathelin could invite Judith Malina and Patrick Stewart to perform. Who knew who would turn up to enliven the show!

It had intrigued me to compare the power structures and belief systems of Fort Worth with those of ancient Kathmandu. Gurkhas and cowboys had a lot in common. No one wants to tangle with them, and they are both great people to hang out with. They both carry themselves with integrity and pride, and have a deep sense of humor. Nearly all Gurkhas and cowboys have a military background. The Stockyards and the Caravan made for a powerful dyad at Fort Worth just as Durbar Square and the sprawling Gurkha barracks did at Kathmandu. IE once held a special week-long conference in those Gurkha barracks. If only the Comanche Indians had been Fort Worth's ceremonial chiefs, it too would have had traditions going back thousands of years.

Ed Bass knew of my intense interest in bringing life back to American city centers. I considered the restoration of a lively cultural life in cities and of farm-based community life to be the two critical needs in American civilization. Reportedly, over twenty people had been murdered in the center of town in Fort Worth in the previous year. A theater and jazz-blues club would bring a genuine, upbeat, safe nightlife into the almost deserted City Center.

Ed, Margaret, and I decided to check out Fort Worth's downtown scene one night before starting this project. It would clearly demand more *savoir-faire* than Las Casas in Puerto Rico, though perhaps less than the Vajra Hotel or the *Heraclitus*. We ventured into a dim, stale-smelling bar for a beer. Two men suddenly took another man into a back alley. They came back sometime later, unaccompanied, with a smile on their faces. The bartender never looked at them. We left soon afterwards, aware that there were objective as well as subjective reasons why Sid Bass had a private police force patrolling his downtown office towers and other properties.

ABOVE: *Winter at Synergia Ranch.* RIGHT: *View from ranch of the Cerrillos Hills.* LEFT: *Library and seminar room.*

Geodesic dome at the ranch is a venue for conferences, workshops and performances.

Over four hundred fruit trees in the Synergia Ranch certified organic orchard produce delicious peaches, apples and pears.

Synopco Construction Projects. TOP: *Project Llano, Palace Avenue, Santa Fe, 1978.* MIDDLE LEFT: *Project Llano.* RIGHT: *Interior of first Synopco construction project, home on Camino Manzano, Santa Fe, 1975.* BELOW: *Two views of Project Tibet, founded in 1980 by Paljor Thondup to assist Tibetan refugees relocation in the U.S.*

ABOVE: Heraclitus *in Phuket, Thailand, 2007.* BELOW: *Life aboard the* Heraclitus; *dinner party on deck; on the Amazon River; in the Antarctic.*

ABOVE: *Hotel Vajra,
Kathmandu, Nepal.*
LEFT: *Tibetan puja
in the Pagoda room.*
MIDDLE AND
BOTTOM RIGHT:
*Hotel Vajra and its
gardens, at the base
of Swayambhunath.*

The October Gallery, London. ABOVE: *Club Room*; MIDDLE: *Café and Gallery*; BOTTOM: *Courtyard and Theater. The October Gallery has, over twenty-five years, successfully established itself as one of London's premier art venues where the finest contemporary art from around the world can be found.*

ABOVE AND TOP RIGHT: *Les Marronniers conference center, Aix-en-Provence, France.*

BOTTOM LEFT: *Trail ride through Birdwood Downs station's improved pastures.* BOTTOM RIGHT: *(top) Cattle in Boab Paddock; (bottom) Bud Crockett helps break in some new Quarab horses.*

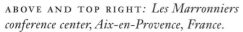

LEFT: *Savannah bungalows at Birdwood Downs offer comfortable overnight accomodation for the Outback traveler.* RIGHT: *Wastewater Garden (in the foreground), a one hundred percent ecological wastewater treatment system at Birdwood Downs station.*

ABOVE: *Some of the wildlife at Las Casas de la Selva, sustainable forestry project, Puerto Rico.* MIDDLE RIGHT: *products made by local artisans from the Blue Mahoe wood.* BOTTOM RIGHT: *Blue Mahoe tree plantations line the slopes.*

ABOVE: *Caravan of Dreams Performing Arts center built by SARBID in downtown Fort Worth, Texas.* BELOW LEFT: *The nightclub and theater.* BOTTOM RIGHT: *Performance at the Caravn of* Cyberspace *by Theater of All Possibilities, 1990.*

A Don Quixote-type motivation soon arose in me, emanating from how I saw Texas power as having the possibility of joining New York and California as pivotal in American politics, and that the Caravan could play a creative role in what others had started to call "The Third Coast." My family had played roles in Texas history. Polk County was named after my grandmother's family; the Walls were ranchers and anti-secessionists in 1861. I was steeped in Texas lore: from Leadbelly and family hero Sam Houston, "The Raven," to Ma and Pa Daniels, to Ed's famous and colorful uncle Sid Richardson. My special interest in Fort Worth came from it being the first big city due south of my grandfather's place, and that it lay on the Chisholm Trail. My grandfather, when leaving the Great Feud, had ridden through it to Oklahoma Territory. Though, as Kathelin and I had discussed, no other theater would settle in Fort Worth. Accepting the risks excited me more than seeing the foolhardiness of this location sobered me.

After much heated discussion of pros and cons, Kathelin and I decided to devote our experience and contacts to the project and Decisions Team agreed to accept Ed's proposal. We made an expedition around London searching for alternatives, but we found that although London might welcome our theater more than Fort Worth, jazz and blues had had a limited impact there, except for the inviting atmosphere that Ronnie Scott had created in Soho. I loved jazz and blues but Kathelin both loved and understood musicians and used the Caravan facilities to make some great recordings, notably of Ornette and his Prime Time group. The first place she had taken me to, in Haight-Ashbury in 1967, had been to hear her musical friends, Jack De Johnette and Keith Jarrett, at the Avalon. For me, that music helped recall my time in 1950-51 organizing for the Packing House workers, when my girlfriends, buddies and I would munch spareribs on Saturday nights, weaving back and forth across State Street on Chicago's South Side.

Jazz and blues are quintessential American musical art forms and Fort Worth had played a keen role in their development, a fact not often mentioned in Chamber of Commerce brochures. Ornette Coleman was the tip of this iceberg and it was his presence that made me decide to undertake the risky venture. From a purely "white" viewpoint, the Caravan of Dreams was impossible in Fort Worth, even though Ed, his younger brother Bob, and the versatile Hyder family showed there was a movement in certain circles. However, with the African-American community's backing, the enterprise was feasible. I was inspired to break the color barrier that still existed downtown and at cultural events in Fort Worth. Not for nothing had my first great artistic teacher been Paul Robeson, and one of my great political teachers, W.E.B. Du Bois in McCarthyite Chicago (1950-52). The slogan of the United Packinghouse Workers Union had been "Black and White, Unite to Fight."

Sophisticated and bohemian people in Dallas would drive the thirty miles to Fort Worth to see something special. Also, the theater could play occasionally in Austin, which already had a lively arts scene. Ed said it meant a lot to him to bring some culture to his hometown. He had become a first class ringer (cowboy) at his Quanbun Downs project in Australia, and this made me think that he could fulfill his dream for his hometown.

The incentive for the DT board was that this project would give us practice in dealing with complex regulations and ideological, artistic, political, and media forces—forces that would obviously come into play when we engaged in a biospheric-scale project. In Fort Worth, these vectors were controlled by a handful of rich but truly civic-minded families, and a few chiefs of powerful corporations at the center of the oil, medical, defense, and electronics industries, like General Dynamics, and Bell Helicopters. Doing the Caravan of Dreams in Fort Worth might serve Decisions Team as basic training had for me in the army. This second joint venture project, more complex than Las Casas, involved investing our entire intellectual capital and moving TAP from Synergia Ranch. It meant giving up our hard-earned position in Santa Fe.

We aimed for the Caravan to be a world-class avant-garde performing arts center, containing a premier jazz and blues club, an acoustically-exquisite Off-Broadway scale theater, and a wonderful meeting place. The exact dimensions of the theater were based on parameters that Kathelin and I had put together from our experience in nearly a hundred theaters and visits to top jazz clubs in New York Village like the Blue Note and Vanguard, and Ronnie Scott's in London. Kathelin knew all the right musicians and I admired them all for their élan, creativity, and lack of airs. Creating a venue devoted to on-the-edge dancers, musicians, and actors also included creating bed-and-breakfast quarters for visiting artists and a theater and arts library. Our concept was designed and built in 1982-83 by Margaret Augustine and engineered by Bill Dempster. This gave us a good start in visualizing a human habitat biome for Biosphere 2 and, of course, lots of legal, political, and logistical experience (such as having to get twenty-three variances to the City Code to build the Caravan). I was quite touched to be made an "Honorary Citizen of Fort Worth" by the Mayor.

The cave bar on the roof came from the time we performed my play *Billy the Kid* in Munich. One afternoon, some of us visited "Good King" Ludwig's Palace. It turned me on with the way it fit perfectly into its site. Between the King's bedroom and the kitchen lay a room carved out of natural rock; a cave in which he could get away from the polished splendor of the rest of the palace. *The Caravan should have such a contrast*, I thought and so did Margaret, so we built a cave for the roof bar – a wonderful contrast to the surrounding steel and glass buildings. We located the waterfall, desert dome, and the

Desert Dome and rooftop bar.

Rooftop view from the Grotto Bar.

Rooftop Grotto Bar so as to create magical spaces, what we called "econiches for conversation." Real conversations did flourish in the gardens on the rooftop and, on special occasions, the atmosphere approached my design criterion, the vibe at the Café de Flore.

Each of our nine projects was now providing us with new skills and introducing us to special people who could help design, build, and operate Biosphere 2, when the moment came. Synergistically, we used the rooftop of the Caravan to build the Desert Dome, and thus integrate IE biome studies with biospheric design studies of Fullerian architecture. With the help of Tony Burgess, a desert ecologist, I put cacti and succulents from four different deserts – some 350 species carefully selected for beauty and ecological interest, representing the Namibian, Sonoran, Malagasy, and Tehuacan Thorn Scrub deserts – in the dome on top of the Caravan. Species from the Northern Chihuahuan desert were planted outside the dome around the rooftop bar. This project enabled us to study world deserts in detail.

Tony became my desert consultant for Biosphere 2. The desert dome also made a great conversation spot for those who attended a play or a music set and wanted to digest their experience beside the roof's waterfall; its white noise sound made the night views of Fort Worth almost otherworldly.

This desert geodesic dome was our first "dry run" at building Biosphere 2. We had made this dome (unlike the one we had built at Synergia Ranch) of glass and metal and at least semi-sealed. In addition, the design of the sound, light, water, botanic, and air-conditioning systems added another layer to Margaret's, Bill's and my expertise. The central architecture, systems engineering, and ecosystem team began to really hum. Meeting the various city regulations gave extra seasoning to both Margaret in her other role as project director, and to Marie as CFO and DI vice-president. Ed managed the finances and politics. Our new consultant for the desert, Tony Burgess, added to his intimate knowledge of the Sonora by looking at the other four world desert systems; he and I would use this new expertise later on when we decided on details for the desert in Biosphere 2.

Together with designing and building the *Heraclitus* and the Vajra Hotel, the Caravan of Dreams gave Bill Dempster, Margaret, and me the confidence to tackle the biospheric super project. I felt our advanced management training program was going well. Ed had proven himself as a financial director.

Kathelin and I wanted to open the Caravan with my friend, master musician Charles Mingus, but he died before the opening. Through the introduction of Kathelin's friend, the style-conscious and magisterial music critic John Rockwell of the *New York Times* (who, like Kathelin, had danced with Anna Halprin), the Caravan opened with my new lifetime friend, Ornette Coleman. Ornette had been born in Fort Worth, one house down "on the other side of the tracks." He had started playing a plastic saxophone underneath the railroad tracks.

IMAGE ABOVE: *Desert ecologist, Tony Burgess, planting the rooftop gardens.*

BELOW: *Plants were selected in the Desert Dome to illustrate the theory of convergent evolution with cacti and succulents from four different desert ecoregions.*

Ornette Coleman at the opening of the Caravan of Dreams. Photo by Brian Blauser. History of Jazz and Blues mural in the background, © Zara Kriegstein.

German prisoners of war from Rommel's army, due to be interned in an alfalfa farm just outside my high school football field, gave Ornette his first realization of the power of his gift. Their train had stopped where Ornette played for practice, and the POWs, though raised on European classics, burst into applause when he finished.

Ornette became an essence friend. He reinforced what I had learned from Robeson and from Kathelin—how to enter the world of music and actually make some myself. I read poetry while Ornette played the sax in a cafe on Greene Street in Soho. I wrote a play in which he and his band, Prime Time, played Orpheus meeting Dionysus—music meeting drama—which, if it had actually taken place, would have changed history for the better. In that packed opening week-end, William Burroughs and Brion Gysin appeared, and I was re-united with three central legends of the 1960s Tangiers School. Of course, I had only been a background player then, soaking in the ambiance created by the masters. Ira Cohen, the great photographer and fellow poet from both Tangiers and the Kathmandu scene, recorded the event.

We opened the Caravan of Dreams in 1983 by showcasing three new works by Ornette Coleman, including a new quintet in honor of Bucky Fuller. It was played by a quartet in the Desert Dome. His major work, "Skies of America," was performed in the Fort Worth Convention Center under the direction of John Giordano. We TAP actors danced in wild costumes down Houston Street from the Caravan to the Center. The Fort Worth symphony alternated with Prime Time in performing sections of the score and improvising. Many of the formally-dressed representatives of Fort Worth's West Side power clique squirmed a bit sitting through this innovative masterpiece.

Film director and editor Shirley Clarke, who had followed Ornette since his big breakthrough in 1959, and Kathelin appeared at every

Shirley Clarke and Ornette in Fort Worth, 1983.

Ornette Coleman, William Burroughs and Brion Gysin, Caravan of Dreams, 1984. Photo by Ira Cohen.

hot moment to shoot footage for a prize-winning biographical film: *Ornette: Made In America*. I served as creative consultant and I will never forget watching Shirley and Kathelin later on in the Chelsea Hotel in Manhattan carefully cutting frames to exactly match the soundtrack. A true work of three great artists: a musician, a film director, and a theater director. The film is still being shown to aficionados. Those hours in the Chelsea marked another step for me in understanding how to be an artist. Ornette and Shirley became two of my great teachers.

William Burroughs and Brion Gysin enlivened the opening with a gravel-voiced, humorous (in Mark Twain's sense) reading followed by polished but uninhibited conversations, reviving old Tangiers memories in me, them, and Ornette. The Jajouka musicians had influenced the four of us. Hamri from Jajouka, Brion's best art student there and the hero of his book, *The Process*, had just shown his work

Gregg Dugan, General Manager of Caravan of Dreams, and Kathelin Hoffman Gray perform Noel Coward's Private Lives *in the Caravan theater.*

at the October Gallery. Brion had opened the pathway to the Jajouka for all of us. We compared the Caravan Rooftop Garden with the Socco Chico in Tangiers and our nightclub to Brion's legendary Thousand and One Nights that closed due to American greed and Berber magic.

The opening of the Caravan of Dreams was covered by the *New York Times* and *Figaro* as a major American cultural event. The *Washington Post* likened the Caravan's opening to "an F-16 landing in an aboriginal village." A tremor ran down my spine reading that. The remark cut to the bone insofar as it was a metaphor for the split between the new and the old in the Fort Worth establishment ... I shuddered because it clearly revealed that the Theater of All Possibilities might be seen as a cargo cult by some and as a danger by others in Metroplex. Its artistic mission might be darkened by stormy weather.

Ted Pillsbury, who ran the Kimbell Museum of Fort Worth with a master's touch, told us to learn from his experience with Fort Worth power people. Kathelin and I had put on a production of our version of Burroughs and Gysin's work *Deconstruction of the Countdown* at the American Center in Paris, run by Ted's brother. We admired Ted's *savoir-faire*. He advised me never to use the word *avant-garde*. The west side of Fort Worth will get you for it. If you do it without the word, they may not notice. The word avant-garde means a threat. Pillsbury elegantly shrugged. Kathelin had scheduled Ornette and William to be followed by the Chicago Art Ensemble, and the Momix Dance Company. I was not going to miss a beat. But Ornette gently and firmly said to us, "You don't know. You really don't know. You should have followed me with a white man; two or three white men."

What excited me was that if the Caravan worked, Fort Worth could become a real creative center and enrich my beloved America with food for the soul. It may have been Ed's home city and a new center in the world market place, but its cattle yards and river and its origin were my home bioregion. So we finished moving TAP from hospitable Santa Fe to the City of Oil, Power, Electronics, and Frontier Legends. Kathelin and I left our less magnificent but heartbreakingly beautiful dome theater at Synergia Ranch, out in the juniper-clad Cerrillos Hills. I had not the slightest notion that Martha Hyder, the intelligent, charismatic, artistic leader of the city, whose presence had been a major factor in persuading Kathelin and me to locate our project in Fort Worth, would soon retire to San Miguel Allende in Mexico. Her departure so soon after our opening was a lightning bolt out of the blue. Nor did we know that our main political friend, Jim Wright, the Speaker of the House of Representatives, would be targeted and destroyed by right-wing ideologists. Sam Houston's Jeffersonian ideals were to be trampled by Texas's rulers once again, as they had done before, in 1861.

Space Biospheres Ventures

CHAPTER EIGHT

This is the space age and we are here to go.
WILLIAM BURROUGHS

IN MY JUDGMENT, THE MOMENT TO LAUNCH BIOSPHERE 2 arrived in May, 1984, when the Caravan of Dreams and Fort Worth had reached a creative equilibrium. We had worked on the science and attracted a cadre of scientists through the Ecotechnic conferences, but Biosphere 2 had to include the viewpoint and understanding of artists. This kind of engineering, modeling the life system of Planet Earth, required a synergy of science and art. It required me to renew my studies of history. Why had this not happened before? What were the obstacles? What were the chances of succeeding?

Besides the importance of eventually building Biosphere 2 to facilitate understanding of our own biosphere, I emphasized the possibility of sending biospheres to other worlds, such as the Moon or Mars, even to a star with the right kind of planet. Although, botanist Tom Lovejoy once exclaimed at one of our scientific meetings, "You *can't* talk about going to a star!"

After the Caravan of Dreams project got underway, I started Space Biospheres Ventures (SBV) as another joint venture company to align activities in two new scientific fields: astronautics, the synergy of physical sciences, and biospherics, the life sciences. Space Biospheres provided a unified field of action. Today, we operate the Biospheric Design Division of Global Ecotechnics, of which I am the chairman. We design laboratory biospheres modeled on the most productive land biomes of Biosphere 2 – tropical rainforests and tropical agriculture – and will eventually make an "Earth to Mars" Biosphere. Our data from Biosphere 2, and our work since 1994 shows that ten thou-

IMAGE ON LEFT: *Biosphere 2, the largest laboratory for global ecology ever built.*

sand square feet (about one thousand square meters or less) can support four earthlings operating exploration technology on Mars for very long periods.

In 1978, when we at Biospheric Design had committed to building a Biosphere, I aimed to prepare to bid for a subcontract from NASA when they began to build the first closed life support system. At that time, many people (and I) still thought that only NASA was capable of building the infrastructure to erect such an innovative, large-scale apparatus. By 1984, it had become clear to everyone that NASA would not build a closed life support system. NASA's scientific policy kept the money flowing to old channels where big profits were made from marketing the discoveries in physical technologies: defense and space shuttle systems and un-manned robotic systems. This reductionist approach to life systems had left the U.S. far behind the Russians.

We were the only outfit – outside of a rumored secret project in Siberia – accumulating intellectual and technical capital to make biospheric systems that included humans. However, if establishment interest ever switched back to Mars, we would be in a very good position to be a major supplier of life systems and to produce spin-off products and programs for recycling. And, if establishment interest broadened because of pressure from environmental crises to find out what was *really* going on with molecular cycles and ecosystems, we would be in a very good position. This gave me reason to believe SBV could attract the venture capital needed from my Manhattan friends and other sources.

Space Biospheres Ventures was owned by Decisions Team (DT) and Decisions Investment (DI) but worked as a consortium with Biospheric Design (BD) and the Institute of Ecotechnics (IE). By 1984 using data from the conferences and our projects, I co-authored a book with Mark Nelson, *Space Biospheres*. It was published in 1986 and translated into Russian. The book met with good response from interested parties and led to my giving talks in Russia, America, Japan and Europe. This was a wonderful opportunity to meet scientists, artists, engineers and supporters of space expeditions whose contributions would be vital to Biosphere 2's success.

My original design had, as step one, a one-person prototype test module at Synergia Ranch. We had a nice little arroyo back in the junipers where the module wouldn't harm the landscape. Step two would be scaling that up to the larger model elsewhere with little risk of our losing control of our ranch test site as our research, development, and engineering center, no matter what happened to the larger enterprise. I figured we needed to raise about five to ten million dollars to make the Test Module. I planned to raise the money by Synergia Ranch contributing the site and infrastructure in exchange for thirty percent of the equity and then selling shares to forward-looking investors who could afford the risk.

I judged that SBV could be capitalized at $15,000,000 since it would provide the project's design and business plan, an integrated management and technical team, the site, and the intellectual capital. Then, we could sell $10,000,000 worth of stock but still keep control. It all seemed feasible, since old acquaintances from Manhattan days had asked me to keep them in mind when I started my next project. The costs, by the way, would be low compared to NASA's due to our constant emphasis on "going for the incremental margins" in every aspect of our work. The six of us on the DT board discussed the pros and cons, after which Ed Bass proposed DI avail itself of its option to initiate Space Biospheres Ventures as a joint venture. He proposed that the joint venture select a site other than Synergia Ranch because the project should start on a bigger scale. Ed suggested DI become the joint venturer by setting up a line of credit. Then SBV could begin with a larger test module prototype, acquire its own long-term site, and get the research and development underway in a fully integrated manner. If the test module worked as well as we thought it would, we could segue to building Biosphere 2.

At this point, Richard Rainwater, Sid Bass's brilliant financial advisor, made a deal with Roy Disney that made Sid a hero. Sid rescued the Disney Corporation from a man whom Roy Disney saw as an arch villain green mailer. Sid and Rainwater installed a decisive new management at Disney, including Eisner and Wells, and gave them *carte blanche*. The new team turned the sick giant around, and the wealth of the four Bass brothers soared upwards.

Ed always kept his main business interests separate from his smaller ventures in ecological businesses and "philecology" donations to the World Wildlife Fund. I had suggested the name because philanthropy without ecology could not produce long-term results. He called this separation the "arms-length doctrine." While our SBV project aimed for a reasonable return on investment, it was still venture capitalized and as such had to be treated with care. Ed sounded sensible and straightforward to me.

The trust that underlay the meetings of the DT board in 1984 is hard to understand without knowing that the DT team had just survived the wildest, most interesting and dangerous theater tour we had ever undertaken. I felt we could not talk biospherics without my friends having a first-hand encounter with Africa; the boundless vitality of that continent had changed my life in 1963-64. So, we played Nigeria at the end of 1983 from Ile Ife and Oshogbo to Zaria and Kano before heading our tightly packed van to the far north to attend the Id-el-Mohammed at Daura, a traditional city near the border with Niger. We survived a military coup, nervous, greedy soldiers shoving rifles in our faces, and thugs disguised as cops stopping us here and there. I would shout (as jovially as possible) "Nigeria! Nigeria! Zaria! Kano!" pointing my finger onward, and the driver would roar forward. Fortunately, no one shot at us.

We met some of Nigeria's greatest living artists, Wole Soyinka and Twins Seven-Seven among them. Taken to meet Wole, we drove out in the countryside where to our horror we saw a parked convertible with one man being beaten by the leader of a gang of thugs. Our driver drove on, to my consternation, but Wole arose, took off his binding ropes, and waved us to come and have a conversation. He had been playacting the role of a crime victim; a none-too-visible camera had been used to attain a sort of grainy realism. We stayed in Twins' palatial compound and fell into the rhythms of the life of a real chief, real artist, and real man. Friends initiated us into the palm wine-drinking club in Ife that Tutuola celebrated in his wonderful novel, *A Palm Wine Drunkard*. That startling, hair-raising, but marvelous Nigerian trip forged new bonds of understanding and friendship among all six of the DT directors.

The DT board approved the deal with DI and then elected me as executive chairman of SBV. They then elected the DT board as the directors of SBV, with Ed as financial director. DT's shareholders also transferred their interests in the Caravan of Dreams to DI as a contribution toward the capital costs of Biosphere 2. That left Kathelin with a free hand as the artistic director of Caravan of Dreams, reporting solely to Ed on that project.

The board commissioned IE's chairman, Mark Nelson, to set up SBV's scientific conferences at the Biosphere 2 site on a budget of $10,000 a year. Members of the board agreed to contribute their labor to build the project. We all continued to donate time to IE, which was run by non-salaried officers. The board comprised about twenty percent of IE members.

Biosphere 2, however, presented tremendous managerial problems … and opportunities. Neither Colorado Mines nor Harvard Business School "conceptual schemes" had evolved to deal with Chaos. I decided to use Chaos theory because of what I had learned about its practicality in previous projects made on the run-up to Biosphere 2. Now periods, or cycles, are attractors which can transform; Ralph Abraham's fundamental definition of Chaos. I modeled our consortium with four Attractors: Decisions Investment represented Command and Control by regular financial monitoring; Decisions Team represented biomic systems (like rainforest, agriculture, waste recycle, etc.) design, creation, and operating; these two corporations formed Space Biospheres Ventures. The third member of the consortium, Biospheric Design, Inc., ran the engineering and architectural groups which gave a unified structure for Biosphere 2 to live in. The fourth member, the Institute of Ecotechnics set up the science meetings from which key scientists volunteered or signed on to assist the the project.

Following the ancient Greek formula, from Chaos emerges Cosmos (order, harmony), this chaotic (dis-orderly) organization with four independent centers of decision, or attractors, worked on my invention or vision of how to make a working model of Earth's biosphere. Finance Attractor could transform from zoom ahead to slow-down to move steadily along; the Synergistic Attractor could transform by increasing, decreasing, or stabilizing the number of components in each subsystem; the Design Attractor could make new plans, tear up old plans, or keep going on the existing plans; and the science could emphasize hypotheses, or naturalist observation, or computer modeling.

I saw my task as inventor, executive chairman, and head of research, and development, to work to make all four Attractors remain independent in an intercommunicating Chaos; therefore each would, as the total system energy built, "struggle" to transform into the most Attractive at any given point, and this gave twelve "Basins" of relative stability, three to each Attractor. Everyone who had any task responsibility therefore rode a continuous learning curve of how to affect the growing reality at every point, which first generated a period or cycle of behavior, which could then transform to a new level of

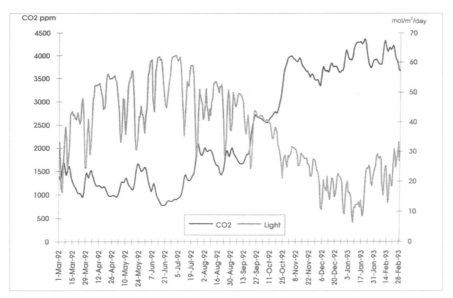

The dance of life and light: A yearly cycle in Biosphere 2 between the incoming light and the amount of carbon dioxide in the air.

understanding. From this Creative Chaos emerged Order, as anyone can see by studying the graph of carbon dioxide and light level fluctuations for the two year period of Mission One, what I call the dance of life and light.

That this Order should develop from Chaos against all their predictions infuriated three classes of conventional managers: those wedded to a cosmos of command and control, those betrothed to a cosmos of falsifiable hypotheses, and those flirts who loved a cosmos of indispensable geniuses.

In short, in the consortium: SBV headed up finance, personnel, research and development; Biospheric Design (BDI) as chief contractor headed up architectural design, construction, and operations; and IE developed contact with scientists. The board had, by consensus, to approve all strategic decisions. I had invented a complex corporate structure to launch one of the furthest voyages yet, the first journey into a second biosphere, one located on this very planet.

Biosphere 2 under construction early 1991; foothills of the Santa Catalina Mountains.

After completing my concept for Biosphere 2 in 1984, I reviewed for myself the critical steps that would have to be taken:

1   *Propose the project to key players and commit myself to work on it until completion.*

2   *Create the joint venture instrument of DT and DI, the Space Biospheres Ventures Corporation, to manage the project in a consortium with Biospheric Design and Institute of Ecotechnics.*

Now I needed to:

3   *Locate a site to build the facilities we needed to do the work.*

4   *Get Biospheric Design to start designing this revolutionary invention.*

5   *Bring together the cadre of scientists, engineers and specialist contractors needed to realize my invention.*

6   *Use the method of progressive approximation: that is, keeping my eye on the total picture, watching how it developed as we reached one and then another critical point where it required direct intervention.*

With points one through two taken care of in early 1984, I turned my attention to three, four, five and six: securing the site, finding the right people, and starting the design process. After considering Florida, California, and certain Caribbean sites (where sunlight would be nearly constant), Marie found a twelve hundred acre site near Tucson, Arizona, which was by far the best of four sites she had discovered during her search. The site had spectacular views of the Santa Catalina Mountains. SBV rapidly acquired it from the Trustees of the University of Arizona.

A small conference center, originally built by Motorola for management meetings, already existed on the site. That center, upgraded and expanded by Margaret, quickly served as SBV's headquarters. Margaret and I calculated it would take another $1,000,000 to make it a fully functioning center for engineering, architecture, and biome studies. That left us $26,000,000 from our $30,000,000 tranche to create and run the Biosphere 2 Test Module and supporting greenhouses. This splendid site, which I called Sunspace Ranch, allowed fairly easy access to the engineering centers of Tucson and Los Angeles. The advent of CADS (Computer Assisted Design) technology permitted us to electronically transmit blueprints. The site's elevation at four thousand feet would help shed the heat load while its southern location would help in getting sufficient sunlight to flux the system. It was not located in the urban counties controlled by Tucson and Phoenix, but rather in one where ranchers and miners were dominant. There was also a penitentiary and several small, established towns. We could get to know everybody in local politics easily.

The theory of biospherics seemed to me to be already well under way with Vernadsky and his geochemical school demonstrating that the biosphere (unknown to most Western scientists at that time) operates as a planetary, even cosmic system. This theory was augmented by Lovelock and Margulis' proofs that bacteria were key vectors in determining atmospheric phenomena, and by Eugene and Howard Odum's studies of mesocosms and the science of ecosystems. My addition was emphasizing the importance of biomic and bioregional systems of flora and fauna, scales above mesocosms like lakes. A scientist named Kamshilov in Russia, it turned out, was working on the essential contribution of biomes and bioregions to the biosphere's long-term stability. I aimed for Biosphere 2 to complete structuring the new science of biospherics and to provide (for the first time) comprehensive data on every vector.

When Marie first showed us around the site, I flashed back some thirty-five years earlier to the winter of 1949 when, at twenty, I picked oranges and cotton and lived in a Farm Labor Supply Camp west of

Phoenix. From that camp, I made my first studies of the Sonoran desert. Its three great features of mountain spine, bajadas (slopes) sectioned by arroyos, and playas (flats) repeated themselves, like Bachian fugues or Homeric episodes, over most of Southern Arizona. This was a bioregion that geologists call the Basin and Range Province. I roamed through snake cactus and thorn scrub in the bajadas with my half-Lakota buddy, Eddie Guthmiller, who liked to hunt for mountain lions with a bow and arrow. He never got one, but the search was something else. One time we heard a lion growling from a nearby thorny brush. We didn't go in!

We fried steaks with bacon strips and cut the taste with blueberries. We'd take a long drink of clear, cool water from our canteens, then lay back while we exchanged stories and watched Orion push the Bull back across the sky. Eddie had known Calamity Jane when she was on her deathbed, and recounted how she had emptied her revolver into her ceiling to pass the time. Then we'd doze off, wrapped in Navajo rugs, speculating on exactly where the Lost Dutchman's gold mine was and how and why so many people had died looking for it.

At Biospheric Design (BD) we busied ourselves extracting design data for Biosphere 2 from the IE conferences and our various bioregional projects. The directors of those projects visited us and we decided how they could help. Las Casas would supply some of our rainforest species; the *Heraclitus* would assist in getting mangroves and the corals, and Birdwood Downs for savannah data. We pored over the literature and had fascinating conversation with visiting experts. We visited others, and engaged top consultants.

Fiery and brilliant Margaret Augustine would serve as project director in charge of architecture, design, and construction, with Phil Hawes as co-architect in charge of the drafting department. I would be ecoscape designer and work with Margaret and Phil on the final

Margaret Augustine, CEO and co-designer of Biosphere 2.

Bill Dempster (on left), Phil Hawes and Margaret Augustine inspect the concrete work on Biosphere 2 construction site.

form the buildings would take; Bill Dempster became the all-important systems engineer playing his ultimate *Go* game: against entropy, corrosion and catastrophe. I would take charge of all research, development, and systems engineering and make sure they all worked together. Margaret and I would have to agree on all points, Bill would have to agree on all engineering points, and Phil would have to agree on the architectural points.

On matters like the all-important structure itself, all four of us worked hard. We brought in the inventive Peter Pearce (Bucky Fuller's greatest student) as consultant. We had to develop the infrastructure needed to create, house, and sustain a world of life comprised of three thousand species. The Earth's biosphere evolved us humans and millions of other marvelous species without which we would all quickly die out, save for a few bacteria which might be able to go into suspended animation, waiting for the next chance. I said that we must remember that not only does form follow function, but function follows form.

Peter Pearce, contractor for the space frame structure of Biosphere 2, with Phil Hawes.

Architecture can transmit cosmogony, as Chartres Cathedral, the Sphinx-Pyramid, the Temple of Heaven, the Palace of Versailles and the City of Palenque prove. What pushed me to work on Biosphere 2's architecture was the vision that our emerging worldview (of quanta, relativity, evolutionary interactions, cosmos, galaxies, solar systems, geosphere, biosphere, technosphere, ethnosphere, cyber-sphere, and noösphere) needed a functional apparatus with a cosmogonic meaning. As with all true cosmogonies, human scale would reconcile fact and value into meaning, with humans this time consciously playing their biospherian role rather than a bio-regional role.

Forms send powerful messages; forms organize the world of appear-ances. Margaret and I hoped to produce an aesthetic arrest that would allow visitors to reach a state of consciousness to slow down, enjoy the forms around them, take the time to soak in the impres-sions and impregnate their imagination. We wished to create spaces that enticed humans to enjoy and expand their inner and outer hori-zons. I wished to create a form that would help induce a future of human habitats harmonizing life, technical, and cultural systems.

We soaked up the insights of architects Bucky Fuller, Frank Lloyd Wright, Hassan Fathy, and Bruce Goff, plus the influences of local biome and cultural creations (as the Chinese junk had influenced the *Heraclitus*). Fuller, Bruce Goff, Hassan Fathy, and Paolo Soleri were my personal friends and Margaret made friends with the last three. Phil Hawes had studied with Wright himself before studying with Goff at University of Oklahoma, where he learned Magic Realism. Our mutual admiration for Goff first brought Phil and me together, and I had carried Wright's *The Future of Architecture* with me wher-ever I went for several years.

Bruce Goff had once invited me to stay with him in his suite at Wright's Price Tower in Bartlesville, Oklahoma. I immersed myself in his drawings while living in Wright's creation. I conversed with a friend of Bruce's, staying in another corner room of that fantastic suite. He was trying to build a Huysmans' piano so that it would play symphonies that gave off whiffs of perfume as well as sounds.

Architect Hassan Fathy lived and worked in a three-story stone house in old Cairo. For a decade and more, each time I found myself in Cairo, some force led me up Old Ottoman Street. My stroll began at a small Egyptian hotel whose entrance faced the hubble-bubbles of El Fashawi and whose four stories of small balconies faced the friendly Mosque Hassan, where some Dervish friends hung out. Nearby stood Hassan Fathy's stone house. I would pull the rope bell inside the flagged courtyard and ascend the stairs.

Margaret and John at the pyramids in Egypt.

Between small cups of heart-warming coffee, Hassan would pull out drawings, unrolling them on his long unpainted table. We leisurely discussed the finer points of adobe for economical and beautiful residences for some Egyptians in the south. I had stayed with a Sufi sheikh where Hassan had built a village near the Valley of the Kings. Hassan always insisted on the importance of context in architecture. He had picked the richest context for his demonstration village! Still, even Hassan had overlooked one thing: the people in an old village of decayed houses he had sought to replace would not evacuate. It turned out that tunnels from those houses led to the graves of pharaohs and nobles and made for profitable plunder!

To develop the vision for Biospheric Design, I took various people to classic architectural sites around the world. We explored Chartres in detail, descending the semi-secret pathway down to the sacred Celtic well located far beneath the still-functioning Romanesque temple to the Black Virgin, which itself is far below the Gothic temple. In Beijing, we walked slowly through the Temple of Heaven; visited the Taj Mahal thrice. Margaret and I both grew to esteem Sinan (the ingenious visionary who built mosques in European Turkey) as a role model for architects and builders. We thoroughly investigated the Pantheon in Rome, the Pont du Gard near Nîmes, Sainte-Chapelle in Paris, Uxmal in the Yucatan, Macchu Picchu and legendary Tiahuanaco (a personal expedition with a Quechuan master) as well as Mameluke Cairo, Karnak and its great volumes of silent stones and calculated spaces, the Pyramids (several visits), and Carnac and its long lines of unsculpted stones.

We strolled through Red Square in Moscow after gazing at St. Basil's and the Kremlin from the windows of the Rossiya Hotel. We appreciated Frank Lloyd Wright's Guggenheim Museum; the reed houses of the swamp people on Schatt al-Arab near Basra, before Saddam Hussein's wanton havoc; the living density of the Jokhan at Lhasa; Khajuraho in Madhya Pradesh and tantric spiritual freedom; the matchless graceful mosques and bridges of Isfahan and other revelations of glory and intelligence.

Margaret and I pondered Ruskin's essays on style. Beauty, the sublime, and the picturesque began to form three central points for me from which all other architectonic values would arise. Beauty's demands can only be appeased by the delight of sensible details, united in wholeness and harmony. Sublimity's demands can only be met by the radiance of complexity – scale and magnitude of varying proportions attuned to the wildness and/or magnificence of its setting. Picturesque's demands can only be satisfied when the free spirit of infinite curiosity is carried ever onward by a romantic matrix of hypertexts interspersed with restful microniches. In Istanbul, one can proceed through the beauty of the Blue Mosque, stand stunned by the sublimity of Hagia Sophia, and wander through the picturesque Topkapi; and view all three from the Bosporus as the sun sets behind them.

Organic Realism became the keynote of Margaret's and my architectural methods as it had Kathelin's and my theater; the emphasis here being spatial rather than temporal. You can no more "strike the set" and move on with a building than you can meander back through interesting parts of a performance and find them still standing there.

Biosphere 2 architecture drew upon ancient and contemporary design influences.

Spaces can enhance the potentiality of human organisms, as, for example, the way you feel your body straightening to its full height when you stand underneath the center of a Gothic ogive arch. While working on the Khuzestan Project in Iran with David Lilienthal in the summer of 1962, I had visited an abandoned mosque in the desert, an hour's drive from Isfahan. Not a single building was left standing near it, but the mosque itself was in very good shape and still being used for special occasions. Standing in its center, my head cocked to take in the dome, I was hurled to the floor by a smack to the back of my head as if from a giant hand. That was my first *ah-ha!* experience of Organic Realism in architecture.

Margaret and I spent several years contemplating these apparatuses that transform the minds and bodies of their participants and transmit history that inspires visitors for millennia, even when they have become abandoned ruins. I don't, of course, deny that some moralizers have laughed at the pyramids as a meaningless joke, a waste of labor, and deplored its use of resources. Even the creator of the Taj Mahal, Shah Jahan, was jailed for the rest of his life by his fundamentalist son, Aurangzeb, for "wasting the treasury for what could have been better spent chastising the infidel." Shah Jahan's architecture has produced centuries of happiness, contemplative thought, and big tourist revenues, as well as giving examples of something better for emperors to do than increasing their armies and persecutions.

Margaret and I alone made the decisions on the final form. We followed the rule that form follows function and function follows form. I thought Frank Lloyd Wright, who emphasized the first half of this rule, did not always properly consider the latter. For architectural contemplations, Margaret and I conferred with Paolo Soleri. Always refreshing and thought provoking, Paolo discussed with us the proper aim of architecture, which is to create a genuine civic life. We added Bucky Fuller's injunction, *Ephemeralize!* Meaning, always try to do more with less, not by cutting quality but by increasing quality and lessening quantity. Our relentless search led us to Peter Pearce. He used his skills in mathematics and visualization so that we did not need any interior supports for our great structures. High-quality engineering is closely allied with harmony, an essential component of beauty.

Some people began to make what they called the Arizona architectural tour in 1991: Taliesin West, Arcosanti, and Biosphere 2. The inner as well as the outer connections are there. I gave talks at both Taliesin West and Arcosanti. I considered Navajo hogans, Walpi, Taliesen West, Paolo Solcri's Platonic city, and Biosphere 2 as five radiant architectures complementing its five geological glories, the Mogollon Rim, the Grand Canyon, the San Francisco Peaks, Black Mesa and the Basins and Ridges.

People appeared at the right time for the right job. Human health care came first! I started our group of advisory scientists with Dr. Roy Walford, an outstanding medical researcher at the University of California at Los Angeles who had seventy people in his laboratories working under National Institute of Health grants. The trim yogic Dr. Roy Walford was a close friend and a true conversationalist who never felt that any topic had been completely explored; that the mystery continued and new insights would arise. I often stayed at his studio in Venice, California. One door was painted white for science, the other painted black for theater. Roy founded and directed his own avant-garde theater group after being turned on by the Situationists, in 1968 Paris. He became well versed in biospherics and, with his rich medical experience, technical virtuosity, and theater, he was the best possible guardian of future biospherians' health.

My aim for the scientific advisors was that these individuals, each with his or her own expertise in research, development, and engi-

Roy Walford at 1982 Institute of Ecotechnics Galactic conference.

neering, would participate in an annual project review for us. Meetings with them reviewed problems that *we* knew we had, and problems *they* knew we had.

I selected other outstanding individuals. I asked Eugene Odum to be an advisor; he was the pioneering systems ecologist whose work on mesocosms and ecology constituted, along with the work of his brother Howard, the closest thing to a guidebook for our design. However, there would be no guidebooks for Biosphere 2 past a certain point because a completely closed set of ecological systems constituted a new level of organization. No scientist ever before had the opportunity to study such a focused complexity! Howard, who strongly supported our efforts, said dryly, "On the basis of previous work, I have to say there's a great possibility you will wind up with a mass of green slime, in a year." The Odums played an essential role.

Rusty Schweickart, an Apollo astronaut who experienced cosmic consciousness on one of his missions, came aboard. Rusty challenged us at every meeting with his famous NASA line: "What are your three biggest problems?" We were joined by Al Hibbs, who calculated NASA Apollo Mission orbits from Cal Tech's NASA Jet Propulsion Laboratory in Pasadena, California and kindly showed me the Galileo spacecraft being prepared for exploring Jupiter's moons. My close friend and supervisor back in 1957, Bob Walsh, the top quality-control expert and thermodynamicist from Allegheny-Ludlum Special Metals, who spent much time educating Japanese corporations on their quality control systems, volunteered. The Theater of All Possibilities spent a hospitable night at his house on one of our tours. Walsh gave Bill Dempster advice on which corrosion-resistant materials were needed to avoid contaminating Biosphere 2.

I asked sharp Gerry Soffen to join after we hit it off at a NASA-sponsored conference. He was so hot about Biosphere 2 that he co-edited

Eugene Odum and Mark contemplate systems ecology inside Biosphere 2.

Howard Odum speaking at the First International Conference on Closed Ecological Systems and Biosphere Science, 1987, organized by IE and held in the chambers of the Royal Society, London.

Rusty Schweickart and Mark Nelson at the first Biosphere conference, 1984.

John with Bob Walsh inspecting the 6XN corrosion resistant stainless steel that they had both worked to develop as metallurgical engineers at Allegheny Ludlum. The stainless steel was used as a liner on top of the concrete throughout the basement of Biosphere 2 and was a key element to accomplish the airtight sealing specifications.

the papers of that conference with Mark Nelson. Gerry was Deputy Administrator at the NASA Goddard Space Center and had been chief scientist for the Viking Mission to Mars. He gave the project his contagious enthusiasm. Gerry compared Biosphere 2 project vibrations with the discoveries and excitement of the Apollo program. Subtle evolutionary geneticist Steve O'Brien at the National Cancer Institute got interested. Studying evolution was an important part of my research plans at Biosphere 2, so I asked him to join. Steve said, "What you're doing is great, new, and unusual. However, you're going to meet some real opposition from the mossbacks. I used to wear long hair. I know how the system reacts to anything different and challenging. You can count on me when the shit hits the fan!"

It was clear to me that something as new and different as Biosphere 2 *would* be threatening to insecure characters in related fields. No one can control or predict what will happen under the immense pressure generated by a large-scale project fueled by ideas and skills that challenge entrenched positions. I introduced a toast at our Sunday dinners: "Remember, anything can happen here!"

I had gained a lot of experience in such matters in my early years at Allegheny Ludlum. The status quo rulers had their gravy train "bread-and-butter" contracts with General Electric and the auto and construction industries. They battled against the introduction of new metals for space and oceanic exploration (especially titanium and zirconium) proffered by my group, led by Dick Simmons and Bob Walsh, as if we were a raiding corporation rather than able employees who could increase their profits and secure their future. During my time at Allegheny, I had been able to watch how Admiral Hyman Rickover dealt with the demanding technical situations that arose with his nuclear submarines, for which I was lucky enough to be tapped to develop zircaloy and beryllium products. I was promoted to the front line of metallurgical creativity by correcting a major mistake caused by my boss and hero metallurgist, Bob Henke. Bob welcomed corrections. A great boss.

I also had invaluable experience with even more intense pressures and unpredictabilities when involved with doing an overall report for Lilienthal on the Iranian Khuzestan Project – the Shah's crown jewel for modernizing Iran. This project, copied from the Tennessee Valley Authority, aimed to green the desert. The project threatened the power of feudal mullahs, whose rule was enforced by beatings and relentless brain-washing. The enlightening effects of this bioregion and community-based project aroused violent antagonisms. The Shah felt forced to throw the Iranian head of the project in jail where Lilienthal and Gordon Clapp of the TVA bravely visited him. As is well known, when Iranian modernizers later split into left and right extremism, ancient hatreds filled the gap, seizing power and using modern techniques to create their dictatorship.

From the standpoint of surviving crises such as those endured by the Khuzestan Water and Power Project, I analyzed my key team, consultants, and biospherian candidates. Biosphere 2's demonstration of the close-knit bond between humanity and all nature would impact at least as many entrenched interests as had the Khuzestan Project and the introduction of Special Metals. I dreamed about these individuals, observed the ways they worked together on our existing projects and how they reacted to pressure or to the lack of pressure.

Most of all I examined *myself*. Undoubtedly, Biosphere 2 would test me to the limit – physically, intellectually, and emotionally. I might think it was all worth doing for this and that historic, scientific, and artistic reason, but what about me? It didn't appeal to me to be put under pressures certain to deliver emotional and intellectual earthquakes, along with inevitable periods in the doldrums. Did I really think Biosphere 2 important enough to sacrifice at least ten years of *me* and a quite happy life in order to get the project up and running when hostility might inevitably stop its completion?

After all, Khuzestan reverted to medieval mullah control and the U.S. space program had been taken over by state bureaucrats. How could Biosphere 2 win? But I knew this inner yapping was intellectual *fol-de-rol*. Nothing could sway me from joining the battle to insert biospherics into human consciousness and history. I felt certain of its necessity and no one else but me looked likely to do it. What had all my arduous studies and risky adventures been for if I didn't use them to the full? No way would it satisfy me to wind up a dilettante.

As dramaturge of the Theater of All Possibilities, I saw two competing scripts that might well simultaneously play as a double screen Biosphere 2: one a mystery epic, the other a tragi-comedy, alternating fairy tale or ambiguous endings.

As a complex systems scientist/inventor, I anticipated intellectual and institutional struggles with the Cold War and the Big Industry, Big University establishment that controlled American science organizations. I could also see biospheric-scale alliances that might occur when open-minded reductionists saw the endless possibilities.

As founder of Space Biospheres Ventures, I foresaw great discoveries and deep friendships emerging from working with naturalists and systems ecologists and live wires from geology and space and energy and geochemistry. Many of them would be hot to check their models of complex systems with the experiment's stream of data from actual occurrences. I also foresaw organized hostility, and personal attacks, the usual destructive opposition to the "new."

As head of research and development, I hoped that cut-and-thrust exchanges would produce sufficient light to see our way forward. I knew that for Biosphere 2 to succeed we had to coordinate the five modalities of science: hypothesis-driven reductionists, discovery-motivated naturalist observers, ingenious physical modelers, iterative mathematical modelers, and pattern-seeking total systems thinkers.

As project entrepreneur, I foresaw an enterprise with long-term pay-offs. These would be generated by the increase in knowledge of the real situation on our planet – what my two friends, Alexander King and Aurelio Peccei, who co-founded the Club of Rome, called "the world problematique." I saw payoffs from onsite income, from our research conferences at this center of biospherics, real-time science tourism, the production of all kinds of closed life system modules, and ecotechnic spin-offs in monitoring systems, ecoscape design, and air and waste recycling modules.

I, as wandering poet-adventurer, foresaw that to escape the tempta-tions and fates of Villon and Rimbaud, I would need to become as clever and quick-witted as Henry Miller and Burroughs, and as at-tuned to ebbs and flows of social relationships as my friend, Lawrence Durrell. His "Alexandria Quartet" might well be simpler than what I would encounter in the Arizona desert. I wondered if Johnny Dolphin, my new *nom de plume et guerre*, could play all the roles required for the different scenes in the drama. Johnny seemed fairly confident, consid-ering all the parts he had played in life, but that confidence might be emanating from his sanguine outlook, not from reality.

As a practical philosopher, I felt accomplished in making neotenous regressions and coming back refreshed and reintegrated (Antaeus touching the earth from time to time to recuperate himself) and in holding together contradictions (Whitman's "Do I contradict my-self? Very well I contradict myself"), and at enjoying happiness in difficult and straitened circumstances. I would have to stay on the *qui vive* to see through phony glamours sure to pop up and try to seduce my attention. And, I undoubtedly needed to work on myself to attain Lilienthal's level of patience and faith in dealing with the fluctuations and hazards of history that must surely strike at least once, like the unthinking violence of an Oklahoma tornado, during the decade needed to design, build, and set Biosphere 2 in motion.

Unless I found someone else who could thoroughly understand biospherics, I would have to make *me* pay a big bill: ten years of unabating attention. At fifty-five, healthy as a horse, I felt I should be able to handle the physical aspect of the task easily, and then turn it over at sixty-five to the partners, whose directors would have reached their primes. This someone else would have to be found and developed because he/she did not yet exist. As for my venture partners, Ed would create a versatile econiche for himself to enjoy his ranches, and gain knowhow for his investment strategies; Margaret could make more great architecture; Kathelin would gain extraordinary material for her dramas; Mark could further develop the Institute of Ecotechnics; Marie would make fine paintings and run the ranch. I myself was determined to finally write my books, which lay scattered in different boxes, aging intimations of possible accomplishment passionately scrawled on dozens of ink-stained, lines-crossed -out yellow tablets, or typed by me on old white paper sheets; visions of Hemingway occasionally arising, stashed in a cardboard box at the bottom of my bookcase.

I caught myself bubbling over with laughter after thinking about all this. All these later projected developments not only depended on my psychologically surviving up and downs, windfalls and pratfalls, knock-downs and drag outs, clamors and silences, but on the continuing consensus and cooperation of the five other highly individualized beings who had set up SBV. Each of them would also be dealing with increasing stresses, strains, and temptations. Each of us already lived in different sectors of society's front lines. I had seen us advancing past all but the farthest scouts; in biospherics, it was we who would, for a time, be society's farthest out, most exposed explorers. And just how receptive the power holders in American society would be to biospherics was an open question. Some would be hostile; there would be knock-down, drag out battles, for sure. I hoped there would not be war. However, the responsibility for getting us engaged was mine. No getting around that.

Nonetheless, Biosphere 2 seemed the greatest adventure around, except for space settlement and, since space settlement can't exist without biospheres, that cut me in on that as well. Not bad for me, essentially a barefoot, blue-jeaned, western Oklahoma frontier-loving boy, no matter how it came out. Rock with my hunches; roll with society's punches; take a break for lunches. My entrepreneurial friends and I had kept the party going for over a decade, four of us for fifteen years, in extreme and far-flung situations. Survivors, one and all. "Anything, as long as *no casualties*" was my invocation. The best chance for us to win was to work together. Our inner team of key players had kept creating and learning under all kinds of circumstances since 1973, some since 1969, with the six of us on the SBV board having trained long and hard for the all-out effort required to pull off Biosphere 2.

We had such good help! There was the resourceful explorer and captain, Robert Hahn, who always landed on his feet – whether in the upper reaches of the Omo River in Ethiopia rescuing our cameraman or repairing a shipwreck in Samoa – searching for sponsors and handling special communications. He had been on the team since 1970. Stylish and sophisticated publisher Deborah Snyder had grown in management stature since meeting us at Les Marronniers in 1982. She had taken over Synergetic Press and would get biospherics out to the public. Calm Chili Hawes piloted the strategic London base for essential meetings with artists, scientists, and biospheric explorers, like Carol Beckwith and Angela Fisher. Chili had been one of our leaders since 1969. Ingenious Robyn Tredwell, who had started with us at the Vajra Hotel project, kept up the rugged savannah systems project in Australia and would fly in from time to time to help us with her expertise on plants. Sparkling Molly Augustine and maestro Cesco Rimondi kept Les Marronniers beautiful for IE conferences. I had great confidence in Roy Walford's impeccable thoroughness in handling the all-important medical side of things.

An intensive two-week course in advanced engineering physiology at the University of Michigan assisted Kathelin and me in understanding the human organism. She would give special help to biospherian candidates in public speaking and improvisation. Talented Phil Hawes would be able to supervise the production of the drawings of the complex architecture that Margaret and I envisioned. Of absolute importance, Bill Dempster, systems engineer of Biospheric Design, possessed the mathematics, drive, experience, devotion to biospherics and patience to decode the toughest engineering conundrums. His contributions to systems engineering would be essential in carrying off the project. He was asked to give papers at the outstanding American engineering societies. Our intense games of *Go* rested both of our minds and prevented us from becoming uselessly obsessed with various problems. If my five board members and these nine held it together working towards our objective, this tested combination of abilities would pull us through any crisis, and enable us to deal as best as possible with catastrophes if they should happen.

In addition, my inner group of four scientific mentors and advisors gave me the confidence that whatever theoretical problems popped up could be handled. This group of widely-acknowledged men consisted of: Earth and Moon geophysicist Keith Runcorn, Fellow of the Royal Society, who, with his deep understanding of and involvement in the overall strategy of "The Campaign of Science," freely contributed insights on how to proceed; Dick Schultes at Harvard, who brought deep wisdom, love, exhaustive knowledge and perception of the rainforest and the ethnosphere; and Howard Odum

Bill Dempster, systems engineer, invented the patented lung technology (variable expansion chambers) for Biosphere 2.

Keith Runcorn, geophysicist and Fellow of the Royal Society, was one of my top scientific advisors.

and Gene Odum, who offered their passion for truth and people as well as their all-important ecological systems expertise.

Keith had helped prove the migration of continents. Dick had founded the science of ethnopharmacobotany. Howard and Gene had practically created systems ecology science, for which they had won the equivalent of the Nobel Prize in ecology. These four made up my personal team of trusted advisors; their honesty inspired me. I knew they would certainly stay the course. In emergencies, I could (and did!) call on these wonderful men; they had seen scientific, political, and cultural life in all its varied forms. They never failed to provide courage, enlightenment, and love. That gave us a total of eighteen people who almost covered the extent of what Biosphere 2 would encounter – pretty good, but not sufficient to get it done.

Richard Evans Schultes, founder of contemporary ethnobotany, Professor Emeritus, Harvard University.

Within a year, I found the four missing persons who would make up my twenty-two colleagues. First, I desperately needed a no-nonsense, practical-minded, theoretical scientist/manager to help me deal with the storm of ecosystem problems revolving around the most important component of biospherics: the microbes. Clair Folsome, microbiologist at University of Hawaii, marathon runner, bare-skinned diver, exobiologist and theorist of life's origins, was just the man. Clair agreed that microbes were the secret creators and maintainers, the Vishnu, of Biosphere 1, Earth. With his friendship, deep interest, and brilliant inputs, we could make sure our microbes worked. That in itself would be a marvelous world; everything else would be gravy.

Second, I needed to solve the waste recycle problem. The Russians had simply handed out the fecal matter after drying it in an oven at their Bios 3 facility. We found Bill Wolverton who ran his ponds at the Stennis Space Center in Mississippi with modest financing from NASA. He was a genuine scientist and real man; theoretical, hands-on, contemplative about implications, friendly to newcomers in his field. He gave us vital data (plus ingenious cues) that finally solved

Wastewater in Biosphere 2 was recycled one hundred percent in these ecological treatment systems, including liquid wastes from the analytical laboratory inside.

how to make a one-hundred percent recycle of human and animal "wastes," the very name of which reflects modern humans' attitude toward a number one fertilizer, essential to recycling, and even the upgrading of life systems. This was, along with ultra-tight sealing and the complete recycling of water and food, one of the four great technical achievements of Biosphere 2.

Third, two of the horrendous biome problems Biosphere 2 faced were the marsh and the coral reef. The *Heraclitus* had taken me into detailed experiencing of these two crucially important biomes. They always gave me moments of what Edward O. Wilson calls "The Naturalist's Trance," where what you see becomes you and you become it; it's not a daze or hallucination, because what you see there translates, with some effort and luck, into detailed understanding. Swashbuckling, likeable, inventive Walter Adey had a preliminary artificial coral reef going at the Smithsonian Institution, and he was deeply

Coral reef collections in Akumal, Mexico, done with permission of the Mexican government and the help of Los Amigos de la Biosfera.

into marshes, especially in the Chesapeake Bay. After stimulating meetings with him and his brilliant, charming wife, Karen Loveland, Margaret and I decided that here was the man to start off with on the ocean design problems.

Walter first built a model of the Chesapeake Bay in the Smithsonian basement. The mangrove biome that resulted from our labors was the most successful of all Biosphere 2 biomes. The Everglades National Park let Walter (and us) collect specimens there on condition that we make all our data accessible. For obtaining the marsh material, we picked up aliquot portions of the chosen marsh's substance – water, soil, slime and small creatures – rather than, as in the continental biomes, a one-by-one selection of specimens. We did select individual mangroves, but each young mangrove was carefully dug and boxed with its surrounding ecology as undisturbed as possible.

Mangroves collected in Florida and trucked to Arizona await planting in Biosphere 2.

Putting the coral reef into Biosphere 2 resulted in a great increase in knowledge about their operation, and in 1992 led to my forming of the Planetary Coral Reef Foundation with Abigail Alling. PCRF used techniques worked out by Phil Dustan, one of America's top coral reef scientists, our consultant at Biosphere 2, to measure and monitor health and vitality of reefs in the Indian, Atlantic, and Pacific Oceans.

PCRF started its series of coral reef expeditions in 1995 led by Christine Handte as expedition chief, and Claus Tober, captain of the Institute of Ecotechnics' research ship, the *Heraclitus*. Christine had captained the *Heraclitus* for several years while it made its home base in Belize, and is now expedition chief of the Planet Water Expeditions voyage on the *Heraclitus* from "Coral Sea to Black Sea."

The ocean coral reef biome was a *tour de force* in the total design; not in terms of needed knowledge gained, but in terms of the effort required and the expense incurred to deal with salt water corrosion. On the other hand, scientifically, aesthetically and experientially, it was marvelous and unique, and some of its methods of assessment were quickly utilized on coral reefs in the World Ocean. Nothing appealed to me more than a swim on that reef or to lie a while in a mild ten-mile-an-hour breeze caressing our beach beneath the rocky cliff leading to the rainforest. A few of those twenty- or thirty-minute intervals in the year before closure date rejuvenated me and my vision of the whole – immensely satisfying moments.

The fourth expert to be found was Ghillean (pronounced Iain) Prance, soon to become Professor Sir Ghillean Prance, Director of Royal Botanic Gardens, Kew. However, Ghillean is seen at his best actually in and amongst forests, tree gardens, and interacting with local cultures where his intuitive rapport with plants, cultures, science, and history makes him incredibly interesting as well as productive. The rainforest biome was (with the coral reef and agriculture) one of our three top biome design problems; Ghillean was the one man in the world who had the knowhow needed to deal with these complexities. At that time, he was vice-president of the New York Botanical Garden and founder of its Institute of Economic Botany, but he spent much time living in the Amazonian forests. Next to Schultes, he was our chief consultant on the Amazon expedition of the *Heraclitus*.

Professor Sir Ghillean Prance, rainforest biome design captain. Sir Ghillean, a master of Amazonian ecosystems, became director of Royal Botanic Gardens, Kew, England. He is currently scientific director for the Eden Project in Cornwall and continues to work on ways to help preserve and protect rainforest biodiversity.

I remarked to Ghillean one time, "Schultes has been initiated into fourteen Amazonian tribes." Ghillean replied, "Yes, I traced his work, and also became initiated in those fourteen tribes." We all admired the work of Bates and Spruce (who began such studies), but I think Dick Schultes and Ghillean became the two great successors and extenders of their achievements. I personally am deeply grateful to them for nominating me as a Fellow of the Linnean Society of London, whose subtle, genial executive secretary, John Marsden, hosted the first international conference on the results of Biosphere 2, in 1996. Without Ghillean, I don't know what would have happened to our rainforest. Margaret managed to get some top plant specimens from Peter Raven, the director of the Missouri Botanical Gardens when they re-did their tropical rainforest exhibit.

With these four, I had twenty-two wonderful people in whom I had the utmost confidence and with whom I could discuss whatever problems came up.

I had systematically tested our board's capabilities by arduous and dangerous expeditions throughout Earth's biosphere. In different combinations, we explored the upper Amazon, Central Asia, Yoruban and Hausan Nigeria, criminal-caste back roads in India, through Tibet from Shigatse to Lhasa, and we survived storms at sea, Outback roundups, and the Saharan vastness. We toured with TAP through historic city streets, splendid theaters, avant-garde back-of-the-wall survival zones, and sacred places. We had been accepted in many cultures and moved in many circles. All of us did depth theater together, had seen glimpses of masks behind masks, even an occasional glimpse of that beyond any mask. We had succeeded in launching progressively more difficult projects.

Vast differences existed in our motivations, education, kinds of friends, family, class, personal temperaments, and individual aims. Each of us had friends outside the team as well as within it. No cultish groupiness can survive the world's shocks. That's why cults retire from the world's actions. These networks of friendships included some of the finest scientists, artists, shamans, and explorers from all regions of the biosphere – a vast reservoir of knowledge about and love for the biosphere. We innovated, on our own as well as collectively. No giant institutions backed us or held us back. I was and am not now hostile to great institutions; I've been in some of them myself and have learned a lot. But our corporations were not beholden to, or puppets of the open market or any unseen string pullers. We had one end to serve: Biosphere 2. Its *Magic If* took me all the way, though finishing it nearly finished me.

Each of us dealt with the world from our own point of view. People drastically changed their behaviors and their values. But at rock bottom, when all the mud was scraped off, each one of us was honorable; we proved it under threat and temptation.

The inner voice – called by some "the wilderness intellect" and by others "the magic mirror" – asked me, "Okay, what have you got to lose?" I replied, "Everything I've done so far." Death would do that for me anyway, so actually I had nothing to lose, infinite meanings to gain. No matter what happened, I would rely on my friends to go forward with or without me. I would not quit on them, hoped they would not quit on me. I aimed past the finish line, the way a Zen butcher aims his cleaver two inches below the cutting board surface. I aimed past completion of the proof of concept of Biosphere 2 in the first two-year mission. I aimed a year afterward, first to the collection, integration, and publications of its results, and then to their embedding in the next mission, come hell, high water, hallelujahs, or all three at once. I pledged to myself never to quit until my mission, establishing biospherics, proved itself in practice, and its theory and history got published no matter how my individual balls bounced along the way.

Many media experts and some powerful reductionist scientists labeled Biosphere 2 impossible. True enough, it was an effort comparable in emotional and intellectual demands, though without the physical danger, to those encountered by the Special Forces unit in Vietnam that I had traveled with as a Stringer Correspondent as they patrolled a large region of armed and wildly different tribal cultures near the Ho Chi Minh Trail. Those of us at the center of SBV had no Fortune 500 Pooh-Bah commanding us, just as those eight master sergeants in Vietnam had no West Point officer commanding them. Each was a specialist in three different military occupational specialties. I had evolved my motto from their excellence and can-do spirit: *Jack of All Trades and Master of Three*. Even my hero, soft-spoken Sergeant Jackson, was only first among equals. Those Special Forces guys were country-wise and technically sharp, trained to perfection, well experienced; they pulled together to carry out their mission and survive. Of course, we didn't have the prospects of ambush to spur us on, but enough creative and moral ambushes lurked to keep us preternaturally alert. We had to keep ready, willing, and able to do our work rapidly, perfectly, and easily.

John (left) with Sergeant Jackson in Vietnam. Jackson's Special Forces were trained to perfection and pulled together to carry out the missions.

CHAPTER NINE

Biospherics

Au fond de l'inconnu pour trouver du nouveau!
(To the depths of the unknown to find the new!)
BAUDELAIRE

AFTER INCORPORATING SBV IN 1984, and after completing my inner organization of twenty-two by the end of 1985, I needed to find and enlist the help of the Russians. Russian scientists had created and nurtured the scientific approach called Cosmism, considering all recurrent phenomena as being part of the Cosmos, not as capable of being reduced to "merely" something or other. This tradition of thought produced Mendeleyev's Periodic Table of the Elements, Dokuchaev's great work on the origins and relationships of soils, Vavilov's discoveries of the main areas where the generation of plant species had occurred, and also Vernadsky's geochemistry culminating in his discovery of the biosphere. The Vernadsky school managed to continue its creative work under Stalin, even though distinguished botanist Vavilov died in a concentration camp, in large part due to his public condemnation by the villainous Lysenko, Director of the Soviet Academy of Sciences.

IMAGE ABOVE: *Bios 3 experiments were run by the Institute of Biophysics in Krasnoyarsk, Siberia, during the '70s and '80s They achieved production of fifty percent food supply, nearly all water and air regeneration and maintained health for three people for up to six months enclosure.*

IMAGE ON LEFT: *View of Biosphere 2 from the rock sculpture on the ridge.*

Outstanding Russians in the Vernadsky tradition had actually begun experimenting with humans in closed life systems: Oleg Gazenko, Ganna Maleshka, and Evgenii Shepelev at Moscow's Institute for Biomedical Problems (in charge of human health in Space, Antarctica, and Ocean high risk areas), and Josef Gitelson and his team, including Nicolai Pechurkin and Lydia Somova, at the Institute of Biophysics at Krasnoyarsk. We had heard the hot rumor that the Russians were doing experiments with people inside a fairly tightly sealed life system for up to several months. The degree of sealing and material recycling was unknown. They had to be Vernadskyans, and

Scientists living and working in Bios 3.

Mark Nelson, left, with Russian scientists, Lydia Samova and Nicholai Pechurkin from the Institute of Biophysics, Krasnoyarsk.

Russian scientist, Galina Nechitailo, who directed life systems experiments on Mir.

they had to know more than anyone else in our field. I had to meet these innovative and versatile Russians. They possessed vital information that I needed to complement the Odums' work on complex ecological systems, the last chapters to finish my guidebook before we soared into the wild blue yonder designing and building a completely sealed complex ecology with one hundred percent waste recycling.

These gifted, deep thinking, idealistic, friendly Russians played a most important role cooperating with us at Biosphere 2. Their experiments had gained invaluable medical results because of Oleg Gazenko's wisdom in the politics of science. It would take years to duplicate the Russians' hard-won knowledge about risks to humans in closed life systems from fungi, bacteria, and trace gas build-ups in atmospheres. Their experience and enthusiasm about the necessity of biospherics being integrated with technospherics and the arts equaled our own. Their great leader, the biogeochemist polymath Vernadsky, had called for the integrating of scientific and artistic thought with political and economic thought to solve the crisis created by the conflict of the technosphere and biosphere. Gitelson said to me that Biosphere 2 *was* Noösphere 1. I replied, "Yes." But biosphere is a notion the American establishment does not cotton on to, and noösphere would make my opponents claim this proved my nuttiness!

The Russians' work differed from ours in that theirs built upon twenty years of state-supported research in actual closed life systems directed toward space applications, whereas ours had been built upon biomic and bioregion work. The Odums had built upon mesocosms like lakes. I unified these three approaches to make Biosphere 2. The Russian research started with Shepelev gallantly existing alone with a chlorella algae reactor (invented by Maleshka), first for a day, then for up to a month. That expanded into Gitelson's three-person study, growing their own crops while in his Bios 3 module for six months. While these experiments had not included waste recycle apart from

Key team of Russian scientific advisors: Josef Gitelson, Oleg Gazenko, Evgenii Shepelev with translator, Leonid Zhurnya, in the Biospheric Research and Development Center.

urine, and Bios 3 had not been able to supply animal protein (which had to be shipped in), they were far beyond what anyone else had done. Of supreme importance in designing Biosphere 2 was knowing that the Bios 3 experiments demonstrated that no fungal, viral, or bacterial diseases (a bugbear to American specialists that we consulted) spread in such a closed environment. The Russians had dozens of doctors watching their projects, doctors who had experience with cosmonauts living in closed artificial atmosphere systems sometimes with a few plants or animals. Without access to this Russian data, Biosphere 2 would hardly have been possible, because starting from scratch would have taken years to prove that sealed-life systems did not pose severe, unacceptable, risks to human health.

In the United States, with no comparable experience with humans in closed life systems, some of our medical and agricultural advisors literally had nightmares about fungus, green slime and bacterial epidemics running wild. In 1986, we built a test module; I volunteered to be the guinea pig and seal myself within it. When I emerged after three days from my "Vertebrate X" experiment as the first human to live in the first one hundred percent water and waste, and ninety percent per year air recycling unit, my medical team rushed me to the checkout room. They feared that my organism would be gripped by a fungus explosion or similar deadly phenomenon. These reductionist fears were based on extrapolations of a single species growth rate without checks or balances in the ecology and without differentiated biomic checks and balances to prevent a single population from exploding. Of course, the secret to my confidence was the advantage of possessing the Russian data and advice, as well as the Odums' ecosystem theory and studies, both of which were highly discounted at that time by American medical specialists.

In 1986, key biospherian players – the six SBV directors, plus Robert Hahn, director of communications, and Deborah Parrish Snyder, secretary to the board – arranged a ten-day visit to Moscow. The central thrust was to meet Gazenko, Shepelev, Maleshka, Gitelson *et al* for comprehensive talks and also to connect with space specialists, poets, moviemakers, and whoever. We evaded the Soviet system of visitor control by taking rooms at the Rossiya Hotel near Red Square, then letting our Intourist guides have the day off after dropping us off at our first appointments. My colorful poet friend, Yevgeny Yevtushenko, who had read in his flamboyant way at the Caravan of Dreams, said, "Why did no one else think of this? To come here to Moscow, sit down for a few days, and make appointments as though you lived here." Someone asked: "How will we recognize him when he gets off the plane at the airport?" Not to worry. Yevtushenko's charisma radiated from the middle of a crowd of disembarking passengers.

We also met, via Gazenko, with Valeri Barsukov and others at the Geochemical Institute founded by Vernadsky. We got to examine in detail the office and library of that great investigator and thinker. The office had been moved into the Institute and preserved exactly as he had left it, including his open copy of *Nature*. But, the Communist regime did not open Vernadsky's thought or his office to the public. Even our Intourist contacts knew nothing about it until Gorbachev's *Glasnost* eventually eased the party's restrictions on allowing access to uncertified thought.

Those were extraordinary days for us in Russia. The unprecedented peaceful revolution grew in force. The Soviet Writers' Union declared its freedom, threw out its Stalinist head, and invited Kathelin and me – because our Ornette Coleman film had won a prize at the Moscow Film Festival in 1985 – to see new uncensored avant-garde films and meet new directors. Andrei Voznesensky read his poems to a packed house of thousands, while thousands more waited outside hoping to get in. He told us that the intelligentsia had made the mistake of opposing Khrushchev because his "thaw" wasn't doing everything perfectly. This time they would back Gorbachev's reforms to the end, even if they disagreed with some of them. Yevtushenko, more ebullient than ever, took us to a Russian Orthodox church crammed with candles and entranced worshippers, where we stood with our backs against the rear wall. Yevtushenko said, "We have to stand here. They don't like people like me from Moscow to go past this point." Then he led us to a secret graveyard in the large cemetery filled with the bodies of soldiers killed in Afghanistan. "The truth about all this must and will come out," he said.

At our request, an Intourist guide took Kathelin and me to see the head of the Stanislavsky Theater. "Who, in your opinion, is the real successor to Stanislavsky?" I asked the director. "It's Michael Chekhov," he replied. The Intourist lady screamed, "Chekhov was a traitor. He went to New York." "Yes," the director shouted back, "because you would have killed him, like Meyerhold!" He leapt to the bookcase and pulled down two Chekhov books in English and waved them in her face. "See, *they* printed his words, but *we* can't print them!"

In 1977 some of us had made our acquaintance with Russian and Soviet ways on a big trip to Leningrad, Moscow, and through Uzbekistan to Samarkand and Bokhara and then on through Afghanistan to the Vajra Hotel. We were on a theater tour that finished with Ben Epperson's and my adaptation of *The Brothers Karamazov* (in order to prepare for Russia) at the prestigious Boat Theater in Copenhagen. It became clear to me on that trip that the Soviet regime was on its last legs. The Leninist bookstores were stripped of Stalinist "classics" and shelves stood bare. The new findings of the Politburo had just been published: they showed that after sixty years, the growth in the working class had topped out at less than fifty percent. Marx's predictions were wrong. Party numbers were declining in the Soviet Union itself, just as they had in the United States and Europe.

The working class basis for Soviet power had been relentlessly eroded by revolutions in technology. We continued on to Bokhara and Samarkand to study the great Sufi (mystic) astronomical and architectural accomplishments in the era of Tamerlane and his successors. We then flew from Tashkent to Kabul to meet some of my old friends (who were to disappear shortly after the Soviet invasion of Afghanistan in 1979). Some Soviet troops were already landing at the airport when we arrived.

We had continued by bus through Pushtunistan, through the Khyber Pass to the bazaars of Peshawar in Pakistan. We met a green-eyed tribal chieftain's son. I asked why he hadn't taken over after his father died. He said, "I would have had to go out, station myself on a mountain trail, and kill a man before becoming chief. I couldn't do that, so I had to leave." After stopping by the Sufi saint Hujwiri's Tomb in Lahore, the "Gateway to India," a bearded dervish, one of those who live "outside the law," ran up to me and took us in hand for two days for a rigorous "training program."

The training consisted of following him through a succession of environments. Two that stood out were: having pebbles thrown at us by students at the university for sitting with this man; and being welcomed with extravagant bows, courtesies, and teas at one of the top restaurants of Lahore. The Sufi idea is that one learns to retain mental-emotional equilibrium through a range of praise-blame rituals that comprise most of a human's daily life. We continued by train to India with a stop at the Sikh's Golden Temple in Amritsar with its great vats of delicious soup for the pilgrims, ending up in Kathmandu, to build the Vajra Hotel Project.

On the 1986 trip, I wanted everyone on our team to become intimate with Russian psychology, history, and culture since their freethinking (though not free to travel) scientists had become a major vector in both biospheric and ethnospheric developments. I also asked everyone to read Engels' *Anti-Dühring* so we could drop the occasional Party-approved quotation when discussing science, and to read Tolstoy's *War and Peace* to get a feel for Russian history. Some of us had worked on a production of Mayakovsky's *The Bedbug*. Producing a play based on Bulgakov's *The Master and Margarita* interested Kathelin and me. Of course, we eagerly discussed Vernadsky's and Tsiolkovsky's ideas on life and space travel and how my notion of Space Biospheres unified their ideas and how we might go about organizing a corporation to realize that vision.

Through Valeri Barsukov, the head of the Geochemical Institute founded by Vernadsky, we also met by telephone the likeable and thoughtful Jim Head from Brown University, an adventurous planetary scientist who played an important role in our thinking about locations of biospheres beyond Earth. Jim led the U.S.-USSR Planetary Geology Committee. Dramatically handsome and overflowing with jovial energy, Barsukov was a master psychologist. He roared in the genial Russian way, "You think I don't know how to get along with Americans? It's four o'clock in the morning in Rhode Island, the worst time in the world to wake a man. I will call Jim up and you will listen to his response. Then you will know we are really friends." Sure enough, Jim emitted a sleepily enthusiastic hello to Barsukov. That scene convinced me Russians and Americans could make real friendships. I was also eager to become friends with Jim; his four in the morning response convinced me that here was a real man who didn't have negative emotions.

Later on, Mark Nelson attended some symposia of Jim's group, and I contributed a short paper on the type of places to land on Mars that might be suitable for a biosphere base. Much of my thinking on the matter had been formed by long conversations with that daring planetary geologist, Jack McCauley, of the USGS Astro-Geology Laboratory in Flagstaff, Arizona, who took astronauts on core desert expeditions to ready them for Moon conditions. Core deserts are the areas where even Bedouins and Aboriginals don't go. Jack and his close colleague, Carol Breed, gave the geology at the Biosphere 2 site a comprehensive walkover, with me carrying the U.S. Geological Survey Map, looking for any fault lines.

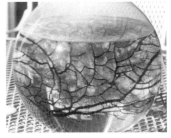

Folsome's microbial studies provided fundamental support to Vernadsky's biosphere theory and the decision to go ahead with Biosphere 2. Shown here is an Ecosphere, a commercial product that emerged from his sealed ecosystem experiments.

We also talked to leading figures in the Russian Space Program. The comparative simplicity of their space vehicle designs intrigued us. These various Russian meetings resulted in free exchanges of important data; personal and institutional alliances were made to advance biospherics and noöspherics that continue to this day.

These meetings brought back memories of me and Margaret visiting that other essential and accomplished experimenter with closed life systems, Clair Folsome. His exobiology laboratory at the University of Hawaii housed his sealed-life systems. Basking in tropical light on his laboratory shelves, Folsome's one- and two-liter ecospheres demonstrated the self-organizing, bacteria-powered capacity of materially closed life systems. These little worlds clearly demonstrated their dynamics. Over periods of time, their basic color changed from red to brown to green, depending on which microbiota became dominant. The oxygen level stabilized after a few months of fluctuations, often at higher than initial levels. After their time at dominance, each microbiota had so degraded their environment that they gave way to a temporarily diminished competitor that bided its time for a comeback. I couldn't help remembering Ibn Khaldun's discovery that Bedouin dynasties lasted only three generations before sinking back into desert obscurity, replaced by stronger Bedouins, uncorrupted by city life.

I had met razor sharp Lynn Margulis to consult about the microworld. Her marvelous book on our microbial ancestors in my library had been underlined and re-underlined. She spoke at one of our IE conferences. Clair's experiments convinced me of the whys and hows of the primary importance and reliability of microbes in building and maintaining biospheres. Complete planetary biospheres existed billennia before fungi, plants, animals, and humans. The Russian Andre Lapo wrote a wonderful book about them, *Traces of Bygone Biospheres*. Those biosphere formations had created our present oxygen-laden biosphere as well as many of Earth's concentrated mineral deposits.

Clair endorsed my idea that one could make sure of microbes doing their task without having to isolate them by simply putting in suites of microbes by using shovelfuls of soil and water from each landscape in each biome. This quick, easy, and economical method of assuring sufficient kinds of microbes in a biosphere works because of their ability to exchange genetic material on a rapid mass basis, creating the ever-changing functional populations needed to handle quickly changing geochemical conditions. (Margulis calls this process an

A collection of Professor Folsome's original flask experiments was displayed at Biosphere 2 Visitor's Center. The small sealed flask systems continued to thrive with microbial life.

"orgy.") Clair said, "From a biospheric aspect, you could consider all bacteria as a single master species." Clair asked Margaret and me to join him at his yacht club on his daily watch for the rare green flash, that marvelous, short optical phenomenon that happens just before sunrise and just after sunset producing a fleeting green spot above the sun. We actually saw one!

Data from Clair Folsome's sealed ecosystems and Russian data on human health in their materially closed life systems gave me the two pieces of knowledge on bacterial systems that I needed to proceed confidently to the finished design of Biosphere 2. Not only did the presence of bacteria not necessarily cause disease, but their resourceful intelligence, beyond that of humans and their technologies, would perform most of the operations in Biosphere 2, just as in Biosphere 1! Clair Folsome, Shepelev, Gazenko and Gitelson all joined my inner circle of scientific confidantes. With the three Russians, my inner circle expanded to twenty-five. With their help, I was ready to deal with anything.

Clair and I concluded that the key to the sustainable prosperity and evolution of Biosphere 2 (and any biosphere) was full suites of bacteria in each patch of soil and each stretch of water. Their orgiastic method of gene exchange ensured the system would be stable by quickly increasing those able to consume buildups in say, methane or toluene. With Clair's analytical backup, Captain Rio Hahn set *Heraclitus*, on the last half of its Round the Tropic World expedition, to collecting bacteria from all its coral reef and marsh stops. Results confirmed our notion about "suites" of microbes. Clair became not only a main scientific advisor, but a loving personal friend. His poignantly unnecessary death in 1989, caused by a fatal reaction to the anaesthetic he received for a minor heart operation, was a tragedy for his wife, for us, for biospherics, and for exobiology.

By 1987 – after the success of the Biosphere 2 Test Module – I felt ready to co-convene the First International Conference on Biospherics, in London. Gazenko from the USSR's Institute of Biomedical Problems and Gitelson from their Institute of Biophysics would participate as co-convenors. Keith Runcorn sponsored our meetings at the Royal Society. SBV and the two Russian Institutes were the only three scientific bodies that had experimented with humans in closed-life systems. We chose London because the Russians couldn't come to the U.S.A., and a lot of westerners couldn't go to Russia. Gazenko said he personally preferred Vienna, but London would be acceptable if the Royal Society would participate. Howard T. Odum represented U.S. systems ecologists. Walter Orr Roberts of NCAR, and NASA geologists and life scientists attended. Ramon Margalef came from Spain. After giving us their thoughtful critiques, these visionary scientists creatively participated in the realization of Biosphere 2 itself.

After that meeting, we started working closely with other Americans best prepared to understand biospherics: Harold Morowitz, the great student of the Krebs cycle, thermodynamicist, and literary stylist, with whom we did a paper on the thermodynamics of biospheric type systems; Walter Orr Roberts, that outstanding student of the atmosphere; Dan Botkin, the systems modeler; and NASA's Frank Salisbury, the brilliant plant physiologist. The Odums (in 1961), and Botkin, Slobodkin, and Morowitz, unbeknownst to me until 1987, had already proposed in 1977 that NASA set up a closed life system with humans. NASA employed reductionist botanists to review their proposals and they had, of course, rejected such wild imaginings. "If we don't understand a plant after a lifetime's work, how can anyone understand a biosphere?" Reductionist ideology straightjackets American science in systems work, to the delight of forces who want no restrictions on their exploitation of the sky, water, soil, and life forms including humans, in their pursuit of profit and power.

CHAPTER TEN

The Biomes

*The stability of the biosphere as a whole, and its ability to
evolve, depend, to a great extent, on the fact that it is a system
of relatively independent biomes.*
M. M. KAMSHILOV

The only source of knowledge is experience.
ALBERT EINSTEIN

IN EARLY 1984, I DECIDED TO MOVE FORWARD with the project with the
first of a series of carefully-planned entrance and exit steps. This
series of step-by-step decision points lowered risks and helped ensure
that experience gained would get integrated into the action. The
other members of Space Biospheres joint venture board approved
the program, elected me executive chairman and we got on with it.

First Entrance: Space Biospheres Ventures buys the property Marie
located; refurbishes the hotel on the site so we can hold engineer-
ing/scientific conferences and live in the old buildings; and, immediately and economically we start doing our research, development,
and engineering. *First Exit:* If that didn't work out, SBV either holds
the site as a property investment or sells it, probably at a good profit.

Second Entrance: If results from step one signal a go-ahead to the
board, we build a Biospheric Research and Development Center
(BRDC) to start the design process with greenhouses to experiment
with the proposed biomes. (I figured we had three years to determine
how the actual biomes would be designed.) *Second Exit:* SBV operates
a new state-of-the-art greenhouse production facility and small luxury hotel, or sells off the property in two or three segments.

Third Entrance: If the BRDC is effective, we design and build a one-
person closed Test Module. Margaret assesses interested contractors

*Test Module built in 1986 to
support one person and to test
many of the basic components
that would become part of
Biosphere 2.*

IMAGE ON LEFT: *Biosphere 2
and the Biospheric Research and
Development complex in the
foreground which housed the
biome experiments from 1986.
These research greenhouses
helped us to select the plants and
animals that would eventually
go into Biosphere 2.*

that we rate as likely candidates on the Biosphere 2 team. With the help of the Test Module, we complete the first approximation of an overall Biosphere 2 design. That finishes our first tranche of money – $30,000,000. We can level off at this point, operating the top planetary research center on closed life systems, plus do cutting edge research in our spacious, open-to-the-atmosphere greenhouse facilities. Each biome has the potential to become a complete biosphere since it possesses all the kingdoms of life working together.

Third Exit: Sale of the facility. At this point, the Test Module might attract but not threaten academia. It could, with some care, be sold to or be contracted out to an existing institution. The greenhouses would interest a major greenhouse producer. Besides supplying a good percentage of the staff's food needs, our fine greenhouse produce would be in big demand in the nearby town of Oracle, Arizona. The greenhouses could easily be converted from experimental use to top of the line vegetable production. Our upgraded conference center could easily be converted to a luxury hotel, offering scenic trails for horse riding.

Fourth Entrance: SBV moves the Biosphere 2 process forward to final size, volume, shape, and design, with new resources coming from the existing joint venture, a new partner, or other venture capitalists. Biosphere 2 is built in three phases: agriculture, wilderness systems, and human habitat. Agriculture first because it is the key system; it has to supply the humans inhabiting this new world with food and delight. In an exit scenario, it would be the world's most sophisticated greenhouse. In our project scenario, it would by itself be by far the most complex closed life system ever operated.

Fifth Entrance: Wilderness (desert, savannah, rainforest, marsh, and coral reef) biomes are finished. They can be run separately but cared for by the same staff running the agricultural biome. Space Biospheres Ventures now possesses the two top facilities for biomic research in the world. It can set up its own educational corporation, or contract out to universities and governments. This pushes the scientific establishment, but is still acceptable to leading reductionists.

Fourth and Fifth Exits: If stopped at the fourth entrance, Biosphere 2 would be the world's largest and most efficient agricultural closed system; if stopped at the fifth entrance, it would be the largest and most measurable wilderness with companion agricultural closed-system and full research center.

Sixth Entrance: Human habitat, the seventh biome, is finished; Biosphere 2 is completed. Mission One starts. The experiment determines whether biospherics science understands the parameters of Earth's biosphere well enough to make a working model. If so, feedback exchanged between biospheres would be of immense benefit. SBV should do well from contracts, spin-off products, publications, and visitor revenues.

Each of these phases would offer unparalleled opportunities to researchers, particularly to us at the Institute of Ecotechnics, Biospheric Design, and, of course, SBV. The sixth phase, Biosphere 2 itself, would offer scientists, artists, and public from around Biosphere 1 an observable ("real time") complex system where humans, their fellow living beings, and "high-tech" interacted to develop parameters for ensuring permanently sustainable co-evolution between life, culture, and technology. If this sixth phase succeeded, a seven-biomed biosphere would come into being, from complex systems science, analogous to Pallas Athena springing forth from the forehead of Zeus.

Sixth Exit: SBV would end the joint venture and decide to stop working on the project, then sell the property as a commercial undertaking, or donate it to a non-profit institution like a university, or sell it off in parts.

The human habitat contained the biospherian apartments, offices, kitchen, dining room, library, exercise room, medical lab, analytical lab and domestic animal bay.

The habitat dining room was a grand space. Materials were all selected for minimum outgassing properties. Carpeting and wall panels were one hundred percent wool, sponsored by the Wool Bureau.

After studying the first five exit points, the board moved Biosphere 2 to its successful history making two-year long Mission One, which proved that biospheric science existed and worked. The decision to proceed to the sixth phase, finishing the human habitat, was made in 1990. (I poked around the underground technosphere of Biosphere 2 looking for Russian-style cot spaces and Spartan desks as a possible minimal human occupied fourth phase if the board decided otherwise.)

My innovative design proposals on biospheric and ecoscape effects had to be approved by Margaret and by Ed. Needless to say, they scrutinized and removed a number of bugs. There was a lot of give and take and push and pull.

The marvelous human habitat biome on which Margaret had lavished all she had learned from Synopco, Vajra Hotel, the Caravan of Dreams and our other projects cost an incremental $10,000,000. That gave the biospherians a proper lifestyle and paperless electronic office so they could carry out multiple duties at peak performance for a two-year period. Margaret's graceful tower soared from the habitat, its library and contemplation space complemented the slightly higher tropical forest biome's straight lines. The architecture expressed a new biosphere-centered cosmogony. It completely satisfied me that our original goal had been reached.

I had four questions for each design point:

1 *Would it help the Biosphere 2 system to increase its initial biomass from fifteen to thirty tons (estimated carrying capacity at maturity)?*

2 *Would it help keep the designed seventy to eighty percent of its initial species diversity, to encourage competition for survival and promote maximum adaptability of the ecosystems?*

3 *Would it help to recycle waste one hundred percent in order to maintain atmosphere and water quality and keep agricultural production high enough and pathogens low enough for Biosphere 2 to remain habitable for one hundred years?*

4 *Would it help to keep carbon dioxide levels below five thousand ppm (parts per million) at the maximum points in the yearly cycle and above one thousand ppm at the minimum points?*

I wanted Biosphere 2 to replicate Biosphere 1 as closely as possible in its state variables (initial forces in the system). One big difference would be that water would exist in only two states, liquid and gaseous. (This happened once in Biosphere 1 and may do so again.) My method of "maximizing difficulties" made Biosphere 2 more a real model of Biosphere 1, which was my basic aim.

Of course, I considered an alternative design for minimizing difficulties, but although beyond anything else done so far, it would not then be an *"experimentum crucis,"* capable of determining the viability of our theories. That route would have led me to select only the most highly productive and resilient rainforest and marsh species, since these produce far more biomass and biodiversity per unit volume than desert, savannah and our shallow ocean could yield. This design approach would have saved immense efforts in gaining comparatively little biomass return from the desert and savannah, and coping with the saltwater required by the coral reef. On the other hand, such an approach would not be anywhere near a model of Biosphere 1. The decision definitely left my head aching from sleepless pondering.

By using maximum artificial light for an agriculture independent of clouds and seasons (offsetting the fifty percent plus loss of light the glass and structural shading produced), I could easily increase agricultural production, but that also meant not replicating Biosphere 1's light cycles. Artificial light's cutting out extra labor *would* allow a cut in the human population down to six or seven and ensure plenty of food.

But I didn't want to miss fully utilizing this marvelous, perhaps once in history chance to do it "as nearly right as possible." I resolved to design in the toughest challenges the system could handle. Biosphere 2 did survive in style and definitively showed that biospheres do immense amounts of self-organizing, including in its humans' actions. It showed that if the technology was rendered non-polluting (minimized pollution) and ephemeralized (maximized information), there could be a synergy between technosphere and biosphere. In other words, it would prove that a noösphere, a sphere of intelligence, *could* exist on Planet Earth if intelligent humans made it happen.

Biosphere 2's success flung down the gauntlet: if it could be done under these difficult conditions, then why not make a noösphere in our much more favored Biosphere 1? It would undoubtedly take a hundred years to arouse that quality of passion, intellect, and vigor in the planet's humans. But, what better, what more necessary, battle to fight?

The jump from biomes to biosphere is gigantic. Biomes are huge, but they are mere chapters in the biospheric book. Stopping at biome studies (as Columbia University later did with Biosphere 2) is a reductionist approach to biospheric studies. In complex systems, several levels of reduction can be made down to a final (in today's science) reduction to the quantum level, but each step "down" marks a further reduction of reality to a conceptual scheme. Biospheric processes – that is, life in materially closed, energetically and informationally open systems – lead to all kinds of physical, chemical, biological, cultural, and physio-psychological interactions. Many vocal critics thought it impossible and nearly all thought it impractical to design, build, and operate an artificial biosphere, especially if it included humans.

I designed Biosphere 2 to sustain a one-hundred-year "human experiment" in order to see how humans deal with relatively long-term life systems, and from that data to lay the basis for genuine noöspherics. So far in 2008, in the fifteen years since Biosphere 2 set world records at the end of Mission One, we can see humans first admiring the splendor of Biosphere 2 and the fact people lived within it; then trying to make scientific reputations and a tourist attraction out of studying its separated wilderness biomes; then putting it up for sale while selling its surrounding area for a scenic housing development. Not that dissimilar from Biosphere 1's history since technological humanity arrived on the scene. Biosphere 2 was a genuine human spectroscope.

Crucial data on long-term biospheric scale atmosphere, hydrosphere, pedosphere (soils), waste, and biomic system events are absent from the world's scientific archives. That's why I started off our biospherian crews with a difficult but vital two-year mission (to establish whether the system was cycling yearly as well as daily). That mission would, if succeeding owners did not bail out, be followed by one-year missions. I doubt anyone else would have opted for a two-year, closed system experiment. Josef Gitelson, who had run Bios 3, told me my greatest mistake, from a political point of view, was not to have run it in three-month periods. Although this "politics of science" approach definitely would have minimized difficulties and greatly eased personal and project criticism, I am unrepentant. Nothing less than a two-year mission showing that the system cycled would have proved that biospheres constitute self-organizing, sustainable entities.

So I called it the way that seemed to me the most certain to accomplish all-important Mission One: only a two year voyage proving the recurrence or non-recurrence of cycles would repay the super efforts that built this time ship. Only a two-year mission could establish details of total system processes inside a biosphere. I knew any mission would meet severe reductionist-led opposition that could easily have shut us down after three months, wasting my and everyone else's time, money, and thought.

I specified five bioregions or "ecoscapes" in each biome to give them the maximum adaptability; it also made them micro-models of reality. The three most difficult biomes to design would be the coral reef, agricultural, and rainforest biomes. Some said I should plant an artificial simulacrum of a rainforest, based on a statistically-determined collection of plants planted on a flat soil, but I decided to model it on the Amazon rainforest, the richest of all rainforests. That meant five bioregions, a cloud forest, a *varzea* or *riverine*, and three relatively drier ecoscapes.

The Rainforest

Ghillean T. Prance, then vice president of science at New York Botanical Gardens and founding director of its Institute of Economic Botany, took a major load off my mind when he agreed to captain the rainforest biome, by far the most important of the three continental wilderness biomes. Rainforests' incredible biodiversity cease-

The rainforest biome seen from the lowland forest up to the cloud forest mountain.

lessly evolves new forms, while its huge standing biomass plays a large role in maintaining the livability (oxygen and carbon dioxide balance) of the atmosphere. Ghillean had the imagination, field experience, and dedication to the rainforest that made me certain of success. He helped train Robyn Tredwell to gather ethnobotanical specimens during our Amazon expedition, so I knew the depth and scope of his love for and knowledge of the rainforest.

I asked him to design our half-acre model rainforest so that if he were standing in the middle of it, he, who had spent years exploring the Amazon, would sense that he was actually in the Amazon. I said, "Let's design the quintessence, not just the essence of the Amazon." Ghillean immediately grasped the concept; his creation beautifully achieved those goals.

Ghillean is a proud, passionate, and conversational man who understands the intricate botanical economics and action metaphysics of the tribal cultures living in the rainforest, as well as the sustainable and co-evolutionary economic uses of its plants. He makes his knowledge and lore available to the planetary public and rainforest peoples. Truly one of a kind, he is at home anywhere – in a remote jungle, at the Linnean and Royal Societies of London or in New York's economic pressure chambers. Modest in all his bearded brilliance, Ghillean acknowledged Richard Evans Schultes as the Grand Master. Ghillean and I, as did Schultes, loved the first great exploring naturalists of the Amazon, the intrepid Richard Spruce and Henry Walter Bates, whose journals had been required reading for our crew on the *Heraclitus*. I considered Ghillean the fourth in this line of great plant explorers.

If the rainforest succeeded, proof of the power of applied biospherics would justify the Biosphere 2 project. It would open the doors to materially-closed biospheric systems because the rainforest and its derivative anthropogenic biome, tropical agriculture, could supply all the food, pharmaceuticals, textiles, and life that would be needed in a biosphere on Mars, the Moon, in the ocean, in Antarctica, or as laboratories and inspirations in world cities. Global Ecotechnics' present-day Earth to Moon and Mars biosphere experiments are based on this design. Schultes called the rainforest the "barely-explored pharmacopeia of humanity."

Its bounty provides the baseline to judge the productivity of all other biomes and is rivaled only by coral reefs. Its importance to the biosphere's overall health, its splendor and mystery, its place as home for spiritually-advanced shamans and cultures, all testify to its incomparable being and power. The brutal rape of sweet-smelling rainforests, accompanied by terror against native peoples, making huge profits for ruthless logging companies, is enough to corrupt politicians in far-off capitals who, for a price, put their police and courts at their service. This horrifying story must be ended once and for all.

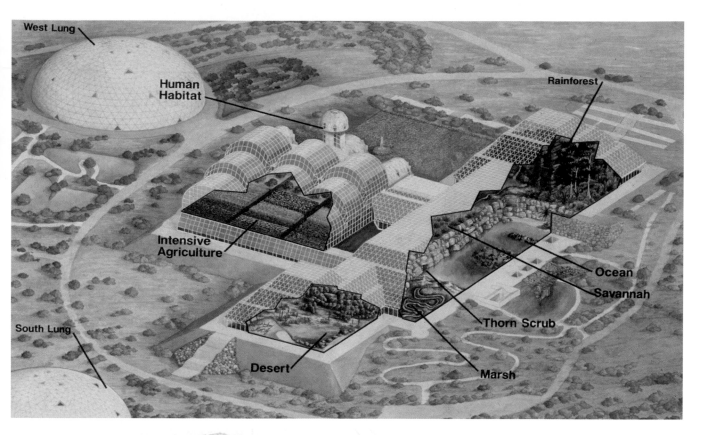

West Lung

Human Habitat

Rainforest

Intensive Agriculture

Ocean

Savannah

Thorn Scrub

South Lung

Desert

Marsh

ABOVE: Overview of Biosphere 2 design.

LEFT: Cross-section of rainforest biome plan.

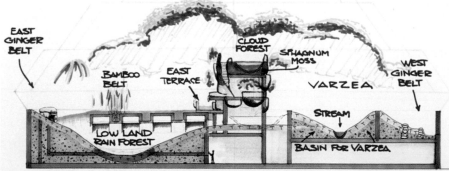

EAST GINGER BELT

CLOUD FOREST

SPHAGNUM MOSS

WEST GINGER BELT

BAMBOO BELT

EAST TERRACE

VARZEA

STREAM

LOW LAND RAIN FOREST

BASIN FOR VARZEA

Co-designer and CEO of SBV, Margaret Augustine, with architect Phil Hawes.

Peter Pearce, founder and head of Pearce Structures, contracted to design and build the spaceframe structure for Biosphere 2.

ABOVE LEFT: *Workers celebrate installing spaceframe on habitat tower.* MIDDLE: *Crews worked day and night to caulk over twelve miles of glass panel seams.* RIGHT: *One of several tests of the stainless steel liner's airtight sealing by flooding it with water.*

ABOVE: *Biosphere 2 under construction in 1990; view into the wilderness biomes from the rainforest mountain. More than two hundred and fifty people helped to build the facility.*

ABOVE LEFT: *The basement housed much of the technosphere of Biosphere 2, including pumps, airhandling units and pipes which moved water (saline or fresh) to each biome. A separate network of closed loop pipes carried heated or chilled water to help regulate the internal temperature and humidity.* RIGHT: *Early experiment in Test Module. Photo by Roger Ressmeyer.*

ABOVE: Biosphere 2, 1991. *"All systems on go." View showing the two lungs in the middle; energy center on the left.*

ABOVE: *Biosphere 2, 1993.*

ABOVE: (clockwise from left) *Desert biome was modeled on the fog desert of Baja California; view of lowland rainforest ecosystem; view from the pool at the end of the upper savannah stream; pathway winding through the lower savannah biome.*

ABOVE: *The intensive agricultural biome was one of the most productive half-acres of farmland in the world.*

ABOVE: (clockwise from left) *Million gallon ocean, modeled on a Caribbean coral reef ecosystem with a shallow lagoon and fore reef; basement rice paddies and various crops; thornscrub biome; marsh biome ecosystems ranging from highly saline mangroves to freshwater wetlands.*

CLOCKWISE FROM LEFT: *Mission Control, the computer center where all data was collected from over one thousand sensors and teleconferencing with the crew inside took place; library in the tower atop the human habitat; dining room and kitchen with window overlooking the intensive agricultural biome.*

LEFT: *Mark Nelson taking soil samples.* MIDDLE: *Mark Van Thillo calibrates one of the sensors in the savannah.* RIGHT: *Sally Silverstone measures out grain from harvest for the chef of the day.*

CLOCKWISE FROM LEFT: *Biospherian Abigail Alling seen through the Ocean Viewing Gallery window tending the ocean; landscaping around Biosphere 2 to accomodate the over quarter of a million visitors each year touring the project from 1991-1993; inside the Ocean Viewing Gallery where visitors learned about Earth's oceans through a variety of exhibits.*

ABOVE: *Visitors outside have an exchange with biospherian, Linda Leigh.* RIGHT: *Russian scientist, Evgenii Shepelev, first man to live in a closed system, looks in.*

CLOCKWISE FROM LEFT: *Bill Dempster in Global Ecotechnics Laboratory Biosphere airlock; exterior of Laboratory Biosphere module; experiments in the chamber with candidate space crops: Sally Silverstone with compact sweet potatoes; super-dwarf wheat.*

CLOCKWISE FROM LEFT: *Wastewater Garden in Legian, Bali at the Sunrise School; a newly planted Wastewater Garden at Emu Creek Aboriginal community in Kununurra, West Australia; coastal Wastewater Garden (in the foreground) at Tropical Padus Resort in Tulum, Mexico.*

The rainforest's silent sublimity meant that, aesthetically, it and the human habitat should take the two high points of Biosphere 2's architectural temple of life. Margaret and I designed the rainforest to be the slightly higher of the two. The forest giants could grow as tall as possible and, artistically or spiritually, the top of their canopies would look down on the soaring column that was the library and contemplation space of the human habitat. Humans at their best need life as well as stars to look up to. The old master himself, Dick Schultes, gave me and our team his apperceptions on this biome. Ghillean's rainforest-loving colleague at the New York Botanical Garden, the ever-positive and adept Michael Balick (a devoted student of Schultes), made important contributions to modeling this most complex of all biomes.

One time while staying with Schultes, Dick said, "You must meet Ed Wilson; he should hear about biospherics." He arranged for me to meet Ed the next day in his roomy Harvard study, from where the great volumes on ants had emerged – fruits of his biophilia – his "love of life" so well captured in his book of the same name. The conversation ranged widely; all seemed to be going well. Ed said, "We need dozens of these biospheres. One on rainforests, one on this and one on that…"

Dick Schultes did a marvelous acting job trying to give me a boost with my distinguished host. I had left my briefcase in his car. Dick came in, looking like an old Indian from the Amazon, head bent down, carrying my suitcase as if he were a native bearer, dripping wet from the rain outside. He set it gently down by my side, then silently left backing out in the same Indian manner.

Deborah Snyder at Synergetic Press published two of Dick Schultes' classic books on Amazon Indians, their forests, and their use of plants. The World Wildlife Fund co-sponsored a meeting at their Washington headquarters to celebrate the publication of Schultes' book,

Where the Gods Reign. Dick looked slowly, knowingly, and majestically around the room and took in its distinguished guests. "I have been," he commenced in his best patrician manner, "invited to speak at every department at Harvard except the Divinity School though I have written two books with 'God' in the title. I console myself with the thought that I..." (he proceeded even more slowly and elegantly) "...know the chemical formula of God." Half the audience looked mystified, a few shocked; the other half permitted themselves a small, knowing smile.

Margaret and I also met with outstanding men at Kew Gardens who agreed to help us, including the lean, resilient, sharp-eyed Martin Sands who had visited our Birdwood Downs experiment in Australia, on a Royal Geographical Society expedition to the Kimberley region. He admired the aged boabs in our paddock, estimating my favorite one to be around fifteen hundred years old. Margaret talked to on-top-of-it-all Peter Raven, director of the Missouri Botanical Garden, who offered some excellent surplus specimens (when they were re-modeling their rainforest). Kew, New York Botanical and Missouri Botanical maintain and nurture three of the most wonderful natural-ist scientific centers on the planet. Every moment I spend in them is magical and profitable to mind and body. They epitomize how public science can and should be run. Visit them if you haven't done so yet.

The Coral Reef Ocean

I got my first insights into coral reefs from our *Heraclitus* expeditions. but for Biosphere 2 design help, Margaret and I turned to that ebullient Smithsonian Institution consultant Walter Adey. Walter's orientation toward mesocosm (a miniature ecosystem open to Earth's atmosphere) and remarkable (but quite small) coral test apparatus at the Smithsonian caused him difficulty in grasping the radical differences between the adaptive behavior of coral reefs in a completely closed system like Biosphere 2 and in one that lay open to Earth's atmosphere.

We worked closely with Walter on the marsh design, which I asked him to model on a Caribbean mangrove biome, after he had run some brilliant experiments on modeling his intimately understood Chesapeake Bay marshes. We loved to spend time with Walter, joining him in contemplating his teeming marsh model, and seeing what new things and events could be seen with this kind of ecosystem observatory. Mesocosm experiments, initiated and developed by the Odums, had indeed been a great first step on the way to Biosphere 2. Walter's preliminary model of the mangrove marsh was a masterpiece of this approach. It furnished a brilliant template for Biosphere 2's design.

After Walter had finished the basic design, Phil Dustan, a slim, intense, coral reef ecologist, who had studied with Jacques Cousteau and the Russians, and who was measuring what happened to coral reefs on the Florida coast, became our top coral consultant. Phil developed a wonderful method of assessing the reefs that we used in Biosphere 2 that is now used on the world's reefs.

I decided to model our coral reef primarily on the magnificent reef system at Turneffe Atoll in Belize, as well as the colorful one at Akumal, up that long limestone Yucatan coast, occasionally hit by fierce hurricanes in Quintana Roo, Mexico. During the building of the *Heraclitus*, Margaret and I had created a "third mind" that could synergize our perceptions and thoughts. We had perfected it in the cultural crucible of Vajra Hotel and she did not shy away from working on this difficult biome design with Walter and me.

Since she was a young girl, Margaret has loved turtles of all kinds. Her big thrill arrived when a giant sea turtle closely circled her,

IMAGE TOP: *Margaret with Gonzalo Arcila at the coral collection site, Akumal, Mexico.*

MIDDLE AND BELOW: *Tiahoga Ruge and her partner, Fernando Ortiz Monasterio, produced and directed a five-part television documentary on Biosphere 2 for Mexican television; together they also co-founded and directed Los Amigos de la Biósphera, a non-profit foundation dedicated to biospheric education in Mexico.*

observing her curiously while she collected coral at Akumal with diver Gonzalo Arcila (with whom I've shared some mind-changing Mayan temple expeditions).

Tiahoga Ruge, one of the U.N.'s World Citizens, in charge of the environmental aspects of NAFTA (North American Free Trade Agreement) as regards Mexico, with ecologist Fernando Ortiz y Monasterio, a fighter for Indian cultures, made a marvelous film about this coral, called *Coral Peregrino* (Pilgrim Coral), shown widely in Mexico. It had been quite a journey from Mexico to Biosphere 2. SBV had designed special small ocean systems with artificial lights and circulating water tanks that fit in the back of sixteen-wheeler trucks so the coral reefs could rapidly and safely be transported from Akumal to the Biosphere 2 reef and ocean. The geology of our reef had to be completed and the water and wave machine operative before the corals could be unloaded by hand and placed in the ocean biome. Any delays on the journey north or malfunctioning of the ocean system on its first trial would mean disaster. A truck-sized coral recycling system has a very short lifetime. To facilitate this delicate operation, the Mexican government had its State Highway Police lead the trucks through difficult traffic areas with flashing lights, saving invaluable hours. Mexico's assistance saved nearly a million dollars in SBV's ocean costs.

In one mountain village, one of the trucks came to a stop. The driver and local mechanic did not know what to do. It seemed a truck part had to be obtained in Mexico City and then sent to the truck. One lost day would drop our safety factor to zero. An old man in the watching crowd made a small signal that Ruge picked up with her director's eye. He pointed to an exact place on the truck. A young man leaped from the crowd, went to the place, removed its covering and fixed some wiring. The truck rolled back in line, losing only an hour. It was neither the first nor last shamanic synchronicity that played a vital and freely-given role in the making of Biosphere 2.

Agriculture biome with small plots of alternating crops supplied year round fresh food.

The Agricultural Biome

When it came to the agricultural biome, I did much of the design work myself. I had been born into a proficient agricultural family and had spent the sixteen years prior to 1985 doing mesocosmic-type agricultural experiments at Synergia Ranch in Santa Fe, Les Marronniers in France, and Birdwood Downs in Australia. My aim was to create an ecologically stable, disease-resistant, tropical agricultural biome – highly productive, health producing, and as easy and interesting as possible to work. I specified a tropical climate for Biosphere 2 agriculture to permit year-round production. It needed at least three full years to develop its complex soil and to adapt to its growing seasons before closure. We studied fifteen hundred different cultivars on site and at the University of Arizona's Environmental Research Laboratory before choosing one hundred fifty hardy, productive ones to supply all the needed ingredients for a human diet.

On the whole, European and American agricultural scientists are devoted to making specific crops more productive by the use of chemicals and altered genetic materials. They are paid by chemical corporations to make quick profits, not to think about developing sustainable ways of farming for individual and tribal farmers whose production of food for themselves never gets added to the Gross National Product.

This *ad absurdum* reductionism, pushed by banks, corporations, and corrupt governments, increases profits by wiping out millions of productive, independent tribal and traditional farmers who live by their own production. Abandoned farms are taken over and turned into factories, using up soil fertility like automobiles use up oil. The United States farming community became the first victim. The U.S. had twenty million independent farmers in 1919; today, there are fewer than two million.

Since the lives of tribal and traditional farmers add up to zero on balance sheets based on cash and credit sales, the profiteering-driven Big Lie says that profit-driven agriculture produces a higher standard of living. It does produce super profits and great power for a few – and devastation for farming communities and farm-based nations. Non-corporate American and European farmers can't survive introduction of these methods supported by immense state subsidies. Their heavy machines, powerful chemicals, and hybrid crops would quickly destroy small farms. They produce natural disasters and lower the health and vitality of life forms even in Biosphere 1's vast spaces. Transporting that system to Mars would be silly.

I did not model Biosphere 2's agriculture on these energy– and culture-wasting factory farms. I combined five different self-sufficient agricultural systems. Each had provided people with plenty of food and plenty of leisure time for millennia. My basic agricultural design idea started with the fact that no existing traditional agricultural diet provides the complete ingredients needed for the maximum health of a human organism.

Making a truly complete diet for humans in Biosphere 2 required synergizing several tropical agricultural systems proved over at least a thousand years: Chinese, Indian, Southeast Asian, Bugandan, and Mayan. A number of expeditions helped me examine these systems and their happy, hospitable ways of life. If three or four of these systems could be grown in the same agriculture, analyses showed that all needed fats, carbohydrates, amines, fibers, and vitamins would be supplied.

We had to train people who were used to thinking in terms of chemicals, machines, or migrant workers to work in this hands-on, ecosystem style of biospheric agriculture. There were two leaders: Sally Silverstone and Jane Poynter. Though born in London, Sally had lived and worked in backcountry villages in Kenya and in Bihar, India which had self-supporting agricultural systems. She had also apprenticed at the Institute of Ecotechnics' Las Casas sustainable rainforest project in Puerto Rico, which included experience with all-important bananas, plantains, and yams. Jane Poynter, also from England, had apprenticed at the Synergia Ranch and Outback Station gardens and animal systems where she mastered seeding, growing, gathering and processing the harvests required for managing this complex one hundred fifty cultivar agriculture system.

I had observed that all great agricultures produced a great, healthy, cuisine. Margaret invited some of the greatest chefs in the world, including Julia Child, to teach their skills to biospherian candidates, who all became good cooks. Margaret herself reached the level of chef by assiduously studying her chefs' demonstrations. My old camping buddy, Bob Bucknell, and I once developed a special steak with bacon on top, sprinkled with blueberries, but scrambled eggs, pancakes, nutritious stews of onions, carrots, unskinned quartered potatoes, and cheap beef are about *my* limit.

Most people are unaware that beef raised in non-agricultural grasslands is tastier and richer in food value than that produced in inhumanly crowded feedlots. Cattle can co-exist with wallabies, or deer, or antelope, depending on location. In Biosphere 2, goats and chickens comprised the animal section and not only supplied eggs, meat, milk, and companionship, but recycled a lot of tough plant material that humans could not digest.

Time had to be figured in for the people who would be sealed in for Mission One to master agricultural procedures. Agriculture, including processing, storing, cooking, and disposal, had to take up less than half of the biospherians' work time so they could perform their other duties and live a well-balanced life. Some agricultural specialists declared that Biosphere 2 would fail because the biospherians would have to work all day just to eat. These well-meaning people had not been raised on working farms like the one my grandfather ran nor lived with traditional farming communities or set up a system like I had at Synergia Ranch. Professionally, they consulted for corporations wishing to increase mass production of a monocrop with the fewest possible (mostly migrant) workers.

These nay-sayers specialized in big machines, high inputs of chemicals and hybrid seeds, and followed the fluctuations of the world commodity market, whereas I had lived in the leisurely richness and happiness of classic agricultural systems like the Balinese, Bugandan and Maya, and recently at our own Synergia Ranch, Les Marronniers, and Birdwood Downs projects. They had no idea of the efficiencies of a good hoe, shovel, rake, axe, adze, hand plow, and water hose. My father grew his incredibly rich garden – that not only fed us but made heaping presents for friends and neighbors – on a fifty by one hundred foot lot on about an hour's work a day in the spring, and a couple of hours on big planting days. I grew up helping with my grandmother's and father's gardens. The dining table was piled high with un-store-bought plenty and variety, whose ease of cultivation left plenty of time for reading, talking, and relaxation, even during the Depression. Of course, there was no TV!

Corporate specialists considered this old-time plenty, eaten mostly by the producers (now, as the ultimate put down, called consumers) with only a portion going to market for sales, as a negligible contribution to their god, the Gross National Product. They also, I thought, underestimated the incentive for biospherians to learn the skills of manual agriculture. The biospherians could not obtain food by flashing dollar bills or credit cards at Safeway. They would only eat well if they learned "Clever-way." If they goofed off, they would eat less. Quick feedback makes learning curves move rapidly upward; I was confident initial production would increase.

A rigorous, visionary scientist, Richard Harwood at Michigan State University, experienced in tropical agriculture, joined our agricultural team as our chief system consultant. Dr. Harwood had assembled a dedicated team of scientists who worked throughout the tropics to increase production without destroying the local systems or degrading the soils. He was one of the few western agriculturists who had completely broken with the notion that industrial agricultural techniques can be imported from temperate zones to the tropic farm systems without driving people off the land and turning the thin soil into red lateritic (iron-aluminum) deserts.

The remarkable Winthrop Rockefeller Winrock Center in Arkansas gave us invaluable advice on the breeds of chickens, milk goats, and pigs to include in our agriculture. However, we needed many hours of detailed research to decide on the final crop cultivars, in spite of the Institute of Ecotechnics' fifteen years of experience with plants. The Environmental Research Laboratory at the University of Arizona had proper facilities to augment our greenhouses to test out the fifteen hundred cultivars. Peer-reviewed scientific papers are

The desert biome design was based on the fog desert ecology of Baja California.

available to assess the success of the agricultural biome. Dr. Roy Walford's papers on biospherian health also attest to the power of a good diet.

The Desert

Tony Burgess, working at the hands-on Desert Laboratory in Tucson, Arizona, under Paul Martin (who proved humans caused the great Paleolithic mammal extinctions in North America), became a pal as well as our desert biome advisor. He zipped around the Sonora like a real desert rat. Tony had done extremely well in Fort Worth setting up the flora in our Caravan of Dreams desert dome, even though these consisted of planted specimens individually looked after. It was not a self-adaptive ecosystem.

As in all the biomes, creating the desert involved using my biospheric design system of levels (starting from the bottom up): microbes, organisms, populations, guilds, biochemical functions, patches, phases, community, ecosystem, landscape, biome, and the biosphere itself.

Tony, at his best, would softly and genially communicate the vivid intricacies of biomes to students or the public. We strolled companionably on some hot afternoon breaks through the majestic saguaro cacti near the Biosphere 2 site. Tony would sweep his hand across a *bajada* or ridge and evoke past climatic epochs like a musician noting chord changes. Margaret helped him as he worked to obtain his doctorate while working to design the Biosphere 2 desert. Tony had a real connection with the desert – that of a poet in love with the form itself.

Tony showed me an almost fairyland biome in Mexico, and I decided to put a Mexican thorn scrub transition between the Biosphere 2 desert and savannah. Its irregular shapes produced an aesthetic touch of romantic mystery. After it was put in, I would hang there for a few minutes on many afternoons to attune myself before we made closure. When I have the chance to tramp through them, thorn scrub bioregions always provoke ecstatic states in me. The one inland from Mombasa in Kenya is my favorite.

Tony Burgess on a collecting expedition for plants destined to go inside Biosphere 2's desert biome.

The Savannah

In 1978, IE initiated Birdwood Downs, a pastoral regeneration project on a five thousand acre property in the Kimberley region of northwestern Australia, with a program of sustainable diversified development including pasture regeneration and production of drought-resistant grass varieties. Being in savannah woodlands (and the Kimberley is archetypal) always refreshes me. Savannahs were to form one of my seven classic type biomes in Biosphere 2. Forested savannahs store up all the varieties of seeds they need to see them through all kinds of changes; some varieties await their chance for decades. I sometimes call it the invisible biome. Savannahs need subtle understanding, so we consulted Tony Burgess for advice.

Peter Warshall, right, with Linda Leigh collecting specimens for Biosphere 2's savannah biome.

A fresh water stream flowed through the savannah biome.

Tony Burgess' sidekick was the intense and versatile Peter Warshall. Peter is an ecologist who can eloquently deliver the King's English and, being so affable a man, gets along with anyone anywhere. Peter helped develop the savannah biome's final composition, expertly combining flora from Guyana, West Africa and West Australia. This blend of flora from different continents I called the "Reunion of Pangaea" (Pangaea is the name for the one-time single-world continent that split apart into separate continents, which resulted in separate continental evolutionary paths). When he over-nighted at Biosphere 2, I would see Peter up and about in early morning, doing his daily quota of bird identifications.

The governments of Venezuela and Guyana gave us support in collecting specimens. Peter, ably aided by Linda Leigh, coordinated the plant acquisitions and plantings in desert, savannah, and rainforest with Tony and Ghillean Prance. These three biome captains had to make sure that their biome's interests did not suffer in the small but vital transition zones; so much evolutionary action takes place there because of the intensity of competition for living space.

George Mueller, the visionary manager of the Apollo program, visited our site anonymously to make sure nobody kept him from evaluating the scene for himself. He caught the spirit of Biosphere 2. Mueller delivered what I really needed to know to lift Biosphere 2 off on its voyage. He is a real American, what Burroughs called a "Jones." Your car breaks down. Jones stops, takes a good look, takes out one of his wrenches, fixes your problem, and drives off with a wave, without even asking for thanks. I adopted two of his Apollo guidelines for getting Mission One going: "All up at once" and "The better is the enemy of the good." Going up all at once meant sealing all the biomes to begin with, not one or two at a time.

"The better is the enemy of the good" meant that I had to get it up and keep it up to prevent perfectionism from draining our resources when something was good enough for achieving its purpose. We won't see perfection in the world of facts. But we won't, can't, settle for less than the good. My insistence on Biosphere 2 staff understanding these two principles made Mission One possible. George Mueller showed me how a great man and philosopher can be humble, observant, common-sensical, thoughtful, and can change your life and project with a couple of sentences. I haven't seen him since, but here is my tribute and my grateful bow. I never think of him without wishing to be more like him.

Unflagging devotion to the Promethean task of creating those biomes caught on and spread among all who worked there. I once heard a construction worker say, "Don't drop that cigarette butt down there. They're making a rainforest." If you've ever seen construction sites, you know what it means! Our work on the biomes proceeded at the same time as the construction, never having to halt either construction or ecological work. The workers and supervisors all knew any qualitative touch would be noted and appreciated. All that construction work went on without any plant being damaged or any animal hurt! Being able to do all lines of work simultaneously cut the costs of the project immensely.

Tucson green home builder and solar pioneer, John Wesley Miller, served as general contractor for all support facilities and acted as energy consultant for the project as a whole. Miller's interest and efforts in the development of an energy efficient, sustainable community led to the creation of the Tucson Solar Village, now known as the Community of Civano, in Tucson, Arizona.

CHAPTER ELEVEN

Engineering

Third in top ten greatest engineering achievements of the twentieth century.

DISCOVERY CHANNEL ABOUT BIOSPHERE 2

No one who has experienced Biosphere 2 can help but marvel at its technological achievement. Despite the enormous complexity of its engineering it functions almost flawlessly. Equally impressive is that eight people were able to support themselves and manage this matter-closed microworld for two full years. These are extremely impressive accomplishments for which the originators can take great pride.

WALLACE BROECKER, LAMONT-DOHERTY EARTH OBSERVATORY OF COLUMBIA UNIVERSITY

I WANTED TO ESTABLISH, ON THE SOUNDEST POSSIBLE BASIS, the science and engineering of biospherics. Fundamentally, new advances go past the limits of the known. To do something new strains your resources to the breaking point but it also arouses the unconquerable spirit of adventure and "the wilderness (meta-cultural) intellect." Staying with the *known* replicates or confirms what has already been accomplished, even if you put the *known* more efficiently through recombinations and extrapolations than anyone else ever has. This property of the *known* is, of course, a good thing, since otherwise we humans would have no expanding base of operations from which to move out into new territories.

Perhaps the best example of my pushing the Biosphere 2 experiment to extremes in order to discover relationships, was deciding not to put in artificial lights for Mission One, even during the difficult winter months when the light inside Biosphere 2 was as low as one-tenth of outdoor summer sunshine in Southern Arizona. In short, I

IMAGE ON LEFT: *Biosphere 2 was the most tightly sealed closed system apparatus ever designed and built - over thirty times more tightly sealed than the Space Shuttle. The white sphere in the foreground was a special patented element called "the Lung," a variable expansion chamber to accommodate the expansion and contraction of the air volume in Biosphere 2 due to temperature changes between day and night.*

chose the location "to push the parameters," though if media and political considerations had been the deciding factors, I would have done it with a complete artificial light system for the agriculture and then coasted home with a "success." Adlai Stevenson's *aperçu* summed up my feelings about that: *Nothing fails like "success."* My scientific ideal is not to acquire data, but to conduct what Newton called *experimentum crucis* – a crucial experiment that so arranges its vectors that its answer eliminates the necessity to do more experiments to prove the result.

Naturally, the more experiments an experimenter can justify, the more funds he gets. But that's like a professor acting as a capitalist rather than as a scientist. I wanted to do for biospherics what Newton had done with the crystal (showing the diffraction of light into seven rays), and what Einstein had done by predicting that in an eclipse you could measure the bending of starlight around a mass (an observation which would make or break his theory).

I wanted all the variables to be focused into one measured action. I wanted to make the Biosphere 2 experiment prove or disprove once and for all my prediction that a biosphere was a highly adaptable, basically self-organizing life system that could even prosper under conditions widely different from actual conditions on Earth. In short, I set out to test Vernadsky's assertion that a biosphere is a cosmic phenomenon and a geological force capable of operating under a broad range of conditions.

Running a biosphere in a temperate zone's seasonally fluctuating light would certainly constitute a good part of such a test. Putting in a guaranteed eighty "moles" per square meter per day of light would not prove anything that NASA and the Russians had not already proved. More light gives more production, up to a certain point. Our range inside Biosphere 2 ran from five to forty moles per square meter per day of light, with an average of about twenty. The

Biosphere 2 Test Module gave me confidence that this adaptability existed, but it had certainly not proved it, due to its small scale.

In 1987, our one-person Test Module succeeded in being the first closed human habitat to attain one hundred percent recycle of waste and water. It had high food production, a record low leak rate of air, and stabilized the daily CO_2 cycle. The system contained so many life forms that it would maintain its own equilibrium, just like the system it came from. Its food production, including an occasional fish, sufficed for up to a month's diet, and the module's system stabilized its carbon dioxide daily cycle during Linda Leigh's twenty one-day enclosure. None of us had health problems or a hint of fungal increase. In fact, the three of us who first enclosed ourselves – me, Alling, and Leigh – all came out radiant and brimming with ideas. The prototype had worked! The board decided not to bail but instead to proceed with Biosphere 2. We had leapt to the top of the field.

To engineer Biosphere 2 we had to solve so many difficult problems. One was the structural integrity needed to cope with all kinds of impacts and stresses on a 3.15 acre footprint complex whose height would go to seventy feet with no supporting columns. Another was the transmission of as much radiant energy as possible through the frame structure, distribution of energy fluxes by winds and tides. Sealing the atmosphere to ten percent leakage a year (less than one percent per month, by far a world record level) so that, for the first time, total atmospheric fluxes in a biosphere could be measured in relation to plant, animal, human, fungal, soil, and water activities. Making sure the summer heat load and winter night radiation did not push the temperatures beyond the design limits. Making sure the changing air pressure-volume (+ or – about thirty percent) did not cause the sealed structure either to implode or explode.

We had to solve completely new kinds of problems, like determining the total volume needed to balance Biosphere 2's atmosphere, water,

Over six thousand glass panels were installed on the space frame structure.

temperature, plant growth, and soils in order to support a population of eight biospherians and numerous animals. Determining the volumes of soils and mucks needed for each biome and bioregion's ecosystems to flourish. Creating a first class medical laboratory that could measure the effects of the Biosphere 2 regime on humans. Maximizing the range and precision of our sensors. Achieving complete water and waste recycling. Producing an artificial intelligence that could operate most of the technical system with the maintenance operator doing it all on half of his or her work time.

Augustine and Dempster made a team. They always somehow managed to do better than I thought possible. Dempster had to review all aspects of the engineering systems. He had total responsibility for the sealing and resulting pressure equilibriums. We had to make sure the ecological criteria were met, by including temperature control, light transmission, water and waste recycling, sealing the structure, efficiency and safety of the operation, the fullest monitoring by sensors, and tracking human health. The blueprints had to be methodically checked for error. The biospheric design work had to meet a strict budget. No government money was involved. Anything given the go-ahead by Dempster in engineering had to be coordinated with BD's architectural review and the resulting design would go to the drafting tables. Those reviewers were Augustine, Hawes, and me. Over twenty years of working together on all kinds of projects with all kinds of people in all kinds of ecologies and political systems made this kind of synergy possible.

No element's operation went unchecked by humans for long. I had learned to coordinate engineering and technics at Allegheny Ludlum with my technical assistants forming a team with the production guys who did the hot rolling, pickling, cold rolling, and so on. I learned even more, running that thousand ton a day crushing and leaching plant at Uravan, Colorado, concentrating uranium and vanadium.

When I left dusty romantic Uravan for the Charles River-watered romantic Harvard University, Bill Gregory sneaked up behind me at the going-away party to ram a carton of mushy ice cream over my head. "One thing for sure," he said, "at Harvard Business School you won't find a buddy that'll do that to you when you start to think you're so damn smart. How will you keep from falling on your ass there?"

Although Bill Dempster reported to me directly on the setting of overall parameters, he was a third and equal member of the structural design team. Either Dempster, Augustine, or I could bring in a key expert. Bill's appraisal flowed with ours, and we would go with that after free-for-all discussion and assiduous measurements – Bill's great forte. Dempster's published papers give a comprehensive picture on how these innovative systems accomplished their task.

From the nearby University of Arizona's Environmental Research Laboratory (ERL), two world-class energy experts and soft-spoken,

John as head of the Crushing and Leaching Plant at Uravan, Colorado, with Elwyn Smith, who took his place when he went to Harvard Business School.

Installing the rubber membrane inside the "lung," a variable expansion chamber designed to deal with air expansion and contraction in Biosphere 2's airtight environment.

intently observant gentlemen, Neal Hicks and John Groh, helped Dempster to calculate the all-important heat loads and cooling curves. We designed Biosphere 2 to run between fifty and ninety five degrees Fahrenheit (ten to thirty five degrees Celsius), a range that allowed us to make economies in power use without risking the health of the plants. John Groh also gave important help on the "lung" design – two expansion/contraction parts of Biosphere 2, which Bill invented and patented. In a closed system, the air pressure inside must be able to stay equal to that outside, otherwise, the structure would explode or implode. There were two lungs; in case of damage to one, the other could carry on until repairs were made. In an experiment that aimed for a hundred-year life, one has to anticipate events that you don't ordinarily face in a ten- or twenty-year experiment.

There was no time for the biospherian in charge of the technical system to average more than half a day on the maintenance of this gigantic facility, if all other tasks were to be accomplished. In order to monitor the enormous number of interrelated events, we had to create a "nervous system" – a five-level, computer-based, data-and-control system. Some technical observers judged it to be the first real Artificial Intelligence.

Augustine secured the assistance of the Hewlett-Packard Company in conversations with Dave Packard. He allowed her to select an elite team to engineer the nerve system whose objectives Dempster, Norberto Alvarez-Romo, an engaging Mexican existentialist and computer expert, and I had laid out. Margaret's deal with Dave Packard ensured the realization of our complex design requirements. We couldn't do Biosphere 2 without the best analytics and communication system. Packard was famous for his "walk around" management technique, and from my first day at Allegheny Ludlum in 1957 that had been my favorite way of keeping up on where high tech action was sizzling or fizzling.

Engineers and accountants work to very close tolerances, and very small margins of error, but ecologists can face thirty percent cyclic fluctuations of populations with assurance. Accountants reduce the most varied vectors to exactly the same money units but engineers and scientists use distinctly different and non-convertible cosmic units of mass, heat, temperature, fluxes, stress and strain, metabolic exchange, and so on. Each specialist had to learn and respect that other specialists on the team had worked out tolerances based on their testing and experience. They could not be dismissed as either nitpickers or slophounds.

Synergizing all these tolerances required "meta-tolerance." Engineers weren't nitpicking when they talked of plus or minus a fraction of one percent. Systems ecologists weren't slophounding when they built in allowances for up to thirty percent extinction rates of species in a given biome before that biome stabilized. Economists weren't arbitrary for insisting that ten percent be tucked away in budget estimates for safety. I had to figure a doable tolerance of tolerances to make final decisions on each biome. If all the tolerances slip-slopped in the same direction and I made meta-tolerance too big, the vessel might flounder in a big storm.

Margaret, Bill and I brainstormed together. We flyspecked all engineering decisions before sending the budget to the finance committee where Ed Bass, the financial director, and Marie Harding, CFO, had to scrutinize it, before Margaret could proceed on actual construction. Fifty people, each representing a body of expert knowledge, worked to create the agriculture biome. About the same number were on the continental wilderness biomes, the human habitat, the technosphere, and the two marine biomes. Some people played on more than one team. To cope with this, besides using formal reports, telephone conversations and literature searches, I expanded Dave Packard's "walk-around" into Johnny's "stroll around and jump in."

The real challenge and excitement occurred when unsuspected vectors and ambiguous relationships came to light and no expert existed because this X or that Y was completely new. That was why research had to proceed hand in hand with development and engineering and why I had to create a complete biospherics conceptual scheme to coordinate them. In this X or Y research, hardcore reductionist approaches could do a necessary job: determine the cause and effect sequence of a single vector. However, results from reductionist studies of a single vector could not simply be transferred to the total system without applying an imperative caveat: *this is true within its limits of error, all other things being equal – which they never are.*

For example, what might be true about a coral reef that was open to Earth's atmospheric fluxes might not be true about a reef open to the greater fluxes of Biosphere 2's atmosphere. Species and individual organisms and systems and multi-systems adapt to changed conditions. One of my favorite scientific principles (which, one may note, is neither a hypothesis nor a theory and has no equation but without which one would never understand chemical systems,) is that of Le Chatelier, which states that if you change a vector in a system at equilibrium, all the other vectors will act to minimize the effects of that change. All causes have effects. Since effects can't escape a closed system, effects in their turn become causes. When this process reaches the system's boundaries, it produces loops and these loops interact. Interacting loops tend to establish stability and cycles of material and information transformation.

Technically, what made Biosphere 2 a biospheric and not an ecological experiment was that we sealed it from the surrounding Earth's air, water, soil, and life. Sealing, whether by physical closure as in Biosphere 2 or by gravitational closure as in Biosphere 1, produces cause and effect sequences. There's no way for a line of causality to escape its effects' energy; this, in turn, activates a new causal sequence that

View of the geodesic dome that will cover the south lung. There were over twelve miles of steel struts and glass panel joints that had to be made airtight.

loops back into the system. This process will continue (cascade) until the system stabilizes or some catastrophe occurs making the system adjust or self-destruct.

Our number one engineering problem was sealing the vast enclosure to my target of a leak rate of one to fifty percent a year, where ten percent, which is what Augustine and Dempster attained, means less than three hundred parts per million (ppm) atmosphere change per day. NASA's closed life systems at Kennedy Space Center leaked five to ten percent a day. Without being close to total atmospheric closure, Biosphere 2 would be merely a colorful appendage to Biosphere 1. Our leak rate was so low that NASA, which checked our Biosphere 2 Test Module out, could not measure it. But we knew there had to be some leakage, so Bill Dempster developed an ingenious method to measure for losses by putting in a known quantity of a rare non-reactive gas, then measuring how much remained in Biosphere 2 over

Biosphere 2 was sealed to less than ten percent per year leak rate, setting a world record in degree of tightness for a closed ecological system. This high degree of closure allowed very precise ecosystem measurements to be made.

time. The Space Shuttle has a leak rate of about three hundred percent a year, thirty times that of Biosphere 2, too much to accurately measure the behavioral interaction of gases like carbon dioxide, methane, nitrous oxide, and ethylene. Russian leak rates have not been officially published but have been estimated to be at one hundred percent or more a year.

If I had made the leak rate target one hundred percent instead of ten percent (which would have still made Biosphere 2 the top apparatus in the world), the experiment would have been much easier to run and would have looked better, but the data loss would have been immeasurable. At this ultra low leak rate, we could measure minute changes in the atmosphere, some in parts per billion.

All inputs and outputs in any part of the closed system can, in principle, be measured and their pathways and flow rates determined because there's no way out. Metabolisms of biospheres can be measured by the rates of change in biomass and in informational, biological, molecular, and atomic components. Just as gravitational physics allows for a unified study of the solar system and the galaxy, so do sealed cycles of life's material exchanges allow for a unified approach to exploring Earth's world of life.

A sealed world of life must be open to receiving sufficient energy flux from an outside source or sources. A biosphere renews and even slightly increases its contained free energy with each cycle, barring such events as big meteor strikes. A biosphere builds its energy mountain by utilizing a small fraction, one percent or so, of the energy passing through it, to upgrade the energy stored in its molecules (i.e., photosynthesis, the making of new molecules using light energy). The efficiency of a biosphere can be precisely measured by the changes in the percentage of energy it extracts and stores from incoming energy.

For example, Biosphere 1 changed the composition of its atmosphere from high carbon dioxide to high oxygen due to the activity of bacteria and plants (which used carbon atoms to make organic compounds, thus making the by-product of immense coal, oil, and gas deposits). This released energy amplified evolutionary fluxes or the "pressure of life," which powered the emergence of nucleated cells, eukaryotes, which evolved into large, freely-moving life forms, like brainy animals and mindful humans. Being much smaller and operating with a turnover cycle of carbon dioxide about two thousand times quicker than Biosphere 1, Biosphere 2 rapidly produced vast amounts of data about patterns of change. We traced fluxes of carbon dioxide from biome to biome, the relation of carbon dioxide changes to ocean pH changes, and further traced those to changes in coral reef vitality.

Biosphere 2, being fully open to information flux visually, orally, and electronically, began conversations, exchanges of information between biospheres that could grow to immense proportions within a century. Energy flux ordinarily results in less energy available to do work (entropy), but a biosphere upgrades free energy (see Morowitz paper in bibliography); information flux can also upgrade a biosphere, that is, contain more information than before. Thus, biospheres can create ever more free energy plus an increase in information. Information feedback between two biospheres with a rise in value I call a cybersphere. A cosmic-level cybersphere thus took its first hesitating steps at Biosphere 2.

The thermodynamics of highly engineered closure are so new to science, so subtly multi-looped, that hardly anyone except American systems ecologists like Howard and Eugene Odum, great plant physiologists like Frank B. Salisbury, great chemists of physiological systems like Harold Morowitz, and Vernadskian Russians could understand what actually was occurring at Biosphere 2. The union of engineering and science in "space biospheres" – synergistically operating outside the consensual, state-of-the-art, and establishment boxes – produces results that provoke some to ignore the data because it requires a change in worldview, which could decrease their political and financial power.

Some highly skilled and experienced people thought our sealing goal was impossible. Even thick concrete leaks at a high rate and the best plastic is porous. I decided to cover the concrete basement foundation with corrosion resistant 6XN stainless steel. We did not seal or epoxy the interior concrete supporting walls and columns since our consultants did not think it necessary. They didn't think the walls would absorb much gas before the system reached equilibrium. That decision turned out to cause long-term difficulties when the concrete *did* absorb about one hundred ppm of carbon dioxide per day, which would hardly have been noticed at a larger atmospheric loss rate.

Eventually we had to replace the oxygen in the carbon dioxide that disappeared in the concrete. The oxygen decline was an unexpected effect and we decided with Roy Walford and the biospherians to "ride the oxygen down" to 14.5% as part of the experiment to see at what level humans could work efficiently. The submarine and mountain climbing and Mars Base people were extremely interested in this experiment and its findings attracted much attention. We learned that concrete would have to be sealed in any future biosphere when used as a liner because it reacts with the carbon dioxide in the atmosphere in such a way that the oxygen in the carbon dioxide becomes sequestered by the concrete.

Mission Two was its own experiment; an experiment in biospherics. Since the American reductionist establishment controlling official and media-backed science doesn't recognize biospherics, some observers claimed it wasn't science. That's why we needed two years of naturalist observation to generate enough data to prove biospherics a science, an independent branch of knowledge about the cosmos.

I had a comprehensive vision that the Mission One experiment would test my hypothesis that an artificial sustainable biosphere could be built, based on biospheric science, and that it would grow within the limits of its resources, stabilize its cycles, and produce information vital to understanding any biosphere, including Earth's. The wondrous thing about a biospheric total system is that no knowledge goes to waste.

The Interacting Systems

CHAPTER TWELVE

Out of many, one. [E Pluribus Unum]
ORIGINAL MOTTO OF THE UNITED STATES OF AMERICA

One for all, and all for one.
ALEXANDRE DUMAS

FOR OUR BIG PICTURE ON ECOSYSTEM DESIGN, Margaret and I first honed in on the basic building blocks, or better, on interconnecting areas of expertise and tradition, or better yet, on the volumes of information derived from interpretations of past and continuing partitions of data related to biospherics. Each of these units would have to be treated as independent but as if in agreement with the rest, as well.

First, I regarded the following major components as semi-autonomous – the seven biomes, the building blocks of the biosphere: the human habitat (micro-city), agriculture, rainforest, savannah, desert, marsh, and coral reef ocean.

Second, the technosphere's five building blocks: the two large lungs that Bill Dempster invented to keep the air pressure in balance with that outside; the power system to supply the heating and cooling accomplished by the water circulating through Biosphere 2 in closed pipes, acting as reversible heat exchangers; the human, animal and plant waste recycle system and technics that must not fail; the nerve center where gigabytes of data would be analyzed to search out meaningful patterns; and the gigantic structure itself, with the minimizing of and recycling of outgassing from its materials.

The third set of components included the four action building blocks: the Test Module; the Biospheric Research and Development

Silke Schneider holds a new-born galago, a small prosimian primate species selected to go in for the two-year experiment.

IMAGE ON LEFT: *Waterfall in the cloud forest ecosystem in the rainforest biome; modeled on the "Lost Worlds" in Guyana, whose government assisted in our plant collections.*

trial biome greenhouses; the Institute of Ecotechnics projects (the Puerto Rican rainforest and Australian savannah, which supplied living material and comparison data for the greenhouse experiments); and the special collecting expeditions to Guyana and Mexico.

Each of these building blocks or strategic areas that Space Biospheres managed had its own team operating its inner integrity and cooperated with other teams for making connectivities. I filled reams of paper with circles, lines, arrows, names, and visualized different permutations of the skeleton that held this huge body together. There were also three supporting building blocks that needed attention: ecoscaping of the grounds; the designing and building of the Visitor's Center; and the creation of staff and consultant's facilities. Margaret and I had to make sure all the blocks worked and evolved synergistically.

We worked from the top down – from the biospheric viewpoint down to the individual microbes or pieces of equipment – and then right back, from bottom up to the biosphere. Each of us kept up with the major players in each of the building block teams, which were split into sub-teams, reporting to the coordinating team. She and I kept ourselves up on one another's work with daily exchanges and frequent in-depth meetings.

After reaching its first approximation, each team used the same method of alternating top-down bottom-up on their selection, acquisition, and integration of the particulars in their building block, or sub-system. I called this way of working "progressive approximation," in which the Institute of Ecotechnics and its predecessor, Biotechnics, had been developing expertise since 1969. Each team had to work in several directions toward whatever other building blocks lay next to their building block, such as developing an ecotone transition between the desert and savannah. I visualized the process as a series of nerve cells connecting to other nerve cells until you can see the brain flashing on and off, here and there, building the capacity for humans to operate from externally- and internally-generated sensings and imagery. I gravitated to where the flashes were most vivid and to where they had disappeared in a blank quiescence. Naturally, each team developed at different time rates due to widely varying logistical problems. For example, the coral reef team could not risk getting coral reef material during hurricane seasons but the agricultural team could get started right away.

Margaret and I integrated ongoing team communications about their situation's progress and problems from the future standpoint of the completed Biosphere 2, including the overall organizing, managing, and evaluating of Mission One in conjunction with the financial, organizational, regulatory, and political facts and developments. These external vectors were reviewed with the board, financial committee, and where necessary with lawyers and accountants.

Each of these vectors changed rapidly so that we always had to maintain an up-to-date picture in our minds. A biosphere is filled with tightly coupled interactions that cascade together. It's not a machine; you can't just bring the separate parts together at the last stop on the assembly line. Margaret added an increasingly necessary new subsystem when she created the publications division under Deborah Snyder, who was wise in the ways of the world in a most charming way. She became Assistant Secretary to the board so she could keep in contact with the thinking on how to proceed to make the ideas and accomplishments available to the public.

While this method of progressive approximation sounds extremely difficult, it works easily, cheaply, and efficiently if there exists a well-tested policy of encouraging all possible intercommunication of self-organizing actions in the system. Studies by Harvard Business School discovered that the most thoroughly planned, buttoned-up-shirt corporations have two organizations, one formal and the other informal.

The informal organization uses progressive approximation, keeps the communication about reality going and makes all the decisions the formal organization leaves out (including some of the most important ones). Thus, new decisions moved out in waves through the whole of our research, development, engineering, and architecture totality, waves distributing energy and information. Margaret and I had to keep a sailor's eye open for the freak waves that occasionally build up and you need to batten down the hatches.

Of course, we not only had to stay as far ahead of all the action as we could, we had to extrapolate and shrewdly guess what would likely happen before it happened. Which team needed to be accelerated or otherwise nudged. We met for two or three hours, twice a week, to run this procedure. In addition, the key site staff of fifteen or so of us all took dinner together except on Saturday nights or for business or scientific dinners with visitors. The central team lived on site. Our commuting costs and distractions were reduced to nearly zero. We lived, breathed, and ate biospherics, while cutting our dining costs by using all the output we could from the agricultural greenhouse.

I was used to living on a work site and eating together, from my days on my grandfather's farm, and from working in the mining industry, in logging camps, in agricultural camps and briefly at a couple of kibbutzes. I had lived in a staff house when I ran the plant at Uravan and enjoyed every minute of it. The houses at Harvard Business School also operated in this manner, as did, naturally, our research ship, the *Heraclitus*. When problems arose, you got a new take on them during dinner's relaxed atmosphere. You got to know everyone very well and learned how to get along when friction arose even if you didn't like it. A few of the more citified live-alone commute-to-office types found this quite a new experience. With the cadre living on the site, it meant there was an always available Biosphere Emergency Response Team (BERT), ready to handle anything.

From Marie Harding's ebullient father, Charlie, a top New York executive, I picked up the practice of holding a Monday morning breakfast with key staff to outline the week's program, difficulties, and objectives. Years before at Charlie's, news came of the Watts riots and the burning down of a section of Los Angeles. Charlie's company had originally financed a number of those destroyed businesses. A disheveled man in a disheveled business suit soon stormed in from the airport. "They've burned down all the businesses in that area; what am I going to do?" His tones were of utter despair. Charlie laid a friendly hand on his shoulder. "Get right back on the plane. Tell them we're extending credit to rebuild – immediately." Tears welled up in my eyes. Essence America in action.

Tuesday nights we invited local artists and intelligentsia and studied critical periods in world history. On Fridays, we followed Rusty Schweickart's suggestion, based on his astronaut experience, and asked ourselves: What are our three big problems this coming week? Occasionally, yes, there were more than three!

On Thursday nights, my old philosopher friend Ben Epperson or I (or occasionally someone else) held forth on the different schools of metaphysics that have influenced worldviews from shamanic times through the present. Egyptian *neters*, Greek philosophers beginning with Heraclitus, Roman Stoics and Epicurean gardens, Hindu gods and worlds, German critiques of pure reason and dialectics, French rationalists and deconstructionists, English empiricists and Newtonians, American Transcendentalists, Idries Shah and Rumi from the Sufis, Naropa's studies of dream, illusion, and lucidity, and Milarepa's songs, Percy Bridgeman's operational view of science, Cohen and Nagel's Logic and Scientific Method, Popper's doctrine on hypotheses, William James, Bohr and Heisenberg, Einstein and Darwin (with Gould's and Eldredge's punctuated equilibrium), Feyerbend's *Against Method*, chaos theory, personally presented by Ralph Abraham,

Gurdjieff's warnings, Lilly's metaprogramming, Vernadsky's discovery of the biosphere, Edward O. Wilson's *Biophilia*, Fuller and his synergetics, and many others.

One of my greatest heroes, Alexander von Humboldt, wrote in his *Cosmos*, "There are two points in the history of man that must not be separated – the consciousness of man's just claim to intellectual freedom, and his long unsatisfied desire of prosecuting discoveries in remote regions of the Earth." It seems to me that separating metaphysics (where intellectual freedom reigns supreme) from operational actualities (where empirical necessity prevails), lies at the root of the pernicious fake religion of scientism and the disastrous fake science of religionism.

We never tried to reach any conclusions on Thursday nights. I don't think anything has ever been "concluded" in metaphysical (philosophical) practice. Synergistic inclusions, without blurring different ideas or viewpoints, without making absolutist conclusions, seem to me the best way to understand this "ecosystem of ideas." If I let myself get carried away with the awesome differences between rainforests, coral reefs, and agricultural biomes in Biosphere 2, I wouldn't see that they all depend on the same atmosphere, gravity, and evolutionary processes. It seems to me that the same approach should be used in living in different metaphysics (worldviews). The technics of metaphysics, like the sciences and arts, constitute a dynamic and permanent modality in the biosphere, technosphere, ethnosphere, and in human individuals.

No biosphere can be understood by ignoring its range of metaphysics, both for their creative power and for their destructive deviations, actual or potential, into *isms* and *schisms* and *what-if-isms*. Earth's biosphere is a cosmic entity in which metaphysics began, evolved, and has played a major role for at least forty thousand years.

Johnny contemplating Biosphere 2 just before closure. All systems on go.

Every culture depends on its metaphysical practitioners, so this irresistible practice, if practiced well, obviously has survival value as well as its delights and hazards. Metaphysics practiced badly can be seen by examining the wreckages of empires and religions whose misleaders pushed some idea to extremes, claimed rights to kill, loot, oppress, and indoctrinate in its name, until they were destroyed by the inchoate rage of the exploited.

Hello to total immersion and the synergy of technical, cultural and life systems. That complex ever-changing work satisfied and inspired me. I felt I was a part of all life, all invention, all human culture. Nearly everyone working with me seemed to share these feelings; work moved along as magically as the *Heraclitus* in full sail on the Indian Ocean.

Prophets of doom popped up, sounding alarms, falling by the wayside only to quickly re-incarnate; their shrill alerts helped me stay awake to hazards. Accidents, psychological causes and effects, type attractions and repulsions, and destiny daily wove a flying carpet of creative imagination. Momentary refreshing stop after momentary refreshing stop, each a small eternity, arose as envoys from brief trances of perfection. These moments nourished me with quietness, even when, on occasion, I had to ask Margaret to lash me to the bare mast while howling back at the storm or pleading to be cut loose to chase some deadly but tempting siren of a notion. Even when far from those refreshing moments, engaged in a meeting where every word had to be watched for tone, denotation, and connotation, an assurance pervaded me that "a place" existed where every action, feeling or thought exists in a lucidity like that just before dawn, when every plant or rock seems to emit its own light. Living was worth it, come hell or high water, and they both came from time to time!

I occasionally took off a weekend to visit the Caravan of Dreams in Fort Worth to assist Kathelin in the production of one of my plays, do a reading of my poetry, or a classic like my adaptation of Goethe's *Faust*, and get her feedback as a board member on the scenes at Sunspace Ranch or take in her original work. We knew very interesting people in Fort Worth who lived their lives in art. Ted Pillsbury ranked with the most thoughtful of museum curators. Stewart and Scott Gentling had become recognized by the Mexicans as experts on Aztec culture. Their exquisite book of paintings illustrating that extraordinary world has a place of honor on my shelves.

And the Hyder family was remarkable! Martha Hyder had rescued several outstanding pianists from their Soviet masters. She organized the Van Cliburn International Piano Competition in Fort Worth, and shone in the artistic circles of San Miguel Allende in Mexico. Brent Hyder was an Oxford scholar and expert on rare manuscripts of Central Asian philosophers. He had worked in Istanbul with Hassan Shushud, a scholarly Sufi acquaintance of mine, on a unique trove of manuscripts from the legendary Khwajagan masters. His father, Elton Hyder, was a connoisseur of thirteenth century Siennese paintings.

Margaret and I visited the inspiring Mike Robinson who was director of the National Zoo in Washington D.C. to gain some insight into animal health, and to pick up on some of his extraordinary observations. He told me about the female octopus who would climb out of her tank at night, move down the corridor to grab an unwary lobster, and steal back to her tank.

As is my custom when in Washington since my days of advanced engineering training at Fort Belvoir during the Korean War, after seeing Mike, I walked to Jefferson's architectural masterpiece, his memorial (which he designed himself) to refresh myself in that vivifying space and to contemplate the sentence he has engraved there. It sums up his life's work: "I have sworn upon the altar of God eternal hostility against all forms of tyranny over the human mind."

To me, the greatest of Jefferson's historic strokes lies in his changing one word of the slogan of Britain's Glorious Revolution, a change that set America on its track of unique greatness. John Locke endorsed "Life, Liberty, and Property," as the three pillars of social integrity. Jefferson undoubtedly contemplated the third word a long, long time. *Life*, yes, of course, although the later disasters of the French and Russian Revolutions that left *life* out of their three power words show that *life* is by no means a higher goal for everyone. Jefferson perceived that the English word *property* allowed for taxes without representation – rent on their North American properties. *Property* led to slavery being defended. Jefferson decided to replace *property* with the *pursuit of happiness*. *Property* was reduced to a rightful place, not an end in itself for humanity.

Back at Sunspace Ranch, I ran my personal American history litany as often before. Jefferson: *Life, Liberty, and the Pursuit of Happiness*.

Jackson: *Bring the people into the White House*. Lincoln: *Of the people, by the people and for the people*. Teddy Roosevelt: *Give them a Square Deal*. Franklin Roosevelt: *Make that a New Deal*. Harry Truman: *A Fair Deal*. Jack Kennedy: *Let's head for the New Frontier*. Then Jack and Bobby and Martin Luther King were assassinated and nobody – including yours truly – dared to raise their head fully in public again. The American ship of state wallows in heavy seas. A TV voice and radiant smile was elected president. Reagan liked Calvin Coolidge's formulation: *The business of America is business*. Then the World Market took over America's leaders: *Entertainment, Business, and the Pursuit of Money*.

Kathelin and I called on our good friend Bob Schwartz in Manhattan, now ensconced in his magisterial apartment overlooking Central Park and walking distance from two of my favorite hangouts, the Natural History Museum and the Frick Museum. I needed someone objective to free-associate with on world trends. This leonine futurist with his magnificent white mane was a management consultant and creator of the Tarrytown Conference Center, which featured an annual conference run by his close friends – Margaret Mead, Joseph Campbell and Judith Crist, the trend-making movie critic of the *New York Times*.

Talking with Bob in 1985 and taking in some of the Manhattan theater scene reminded me that Kathelin's and my main sacrifice making Biosphere 2 was giving up our annual theater tour around Europe and North America. Our last one, as it turned out, had been in the fall of the previous year, 1984, when TAP's fantastic production of my *Gilgamesh* script culminated our North American tour, from New York to Vancouver and down the Pacific Coast. We used a one hundred-acre "stage," winding through the forest glades just below the timberline for our finale on Mount Popocatépetl in Mexico.

We performed at the invitation of the Mexico City Psychoanalytic Association. Some associates of Claudio Naranjo had been intrigued by our presentation of the archetypal characters, some which reincarnated a millennium or two later in characters like Odysseus and Noah. It took three days to acclimatize to the nearly three-mile-high elevation and do Kathelin's trademark dance and fight scenes without panting. At one point Gregg Dugan (who played Gilgamesh) jumped off a twelve-foot cliff onto a sandy arroyo.

In late 1986, I suggested to the board that I should resign my post of executive chairman and that its duties be split between Margaret as CEO and Ed as chairman, in order for me to concentrate on the rapidly growing research, development, and engineering needs of Biosphere 2. The accelerating workload of synergizing the three wild horses of research, development, and engineering took all my time; something had to go, and fortunately Margaret and Ed could take over these two management posts.

Never-done-before things were piling up even in the hands of my incredibly dedicated and inspired teams. Eight people had to be able to live healthily, happily, and creatively in this biosphere; its design lifetime: one century (but of course that would depend on ownership-management continuity). Its biomass needed to increase from fifteen to sixty tons in ten years, its seven biomes flourish, its technosphere operate efficiently without toxic buildups in waste, water, air, or food recycling, and the daily atmospheric cycles must oscillate seasonally at life-friendly levels. Its cuisine had to supply all the needed elements, in delicious form.

Its biodiversity should stabilize at about two thousand plant, animal, and fungi species. The biospherians would have to produce their scientific papers and participate in educational, technical, and ecological meetings from time to time. The data that poured out from the ap-

One of a thousand sensors placed throughout Biosphere 2 to monitor environmental conditions including temperature, wind speed, light, oxygen, carbon dioxide, trace gases and other key variables.

proximately one thousand sensors had to be integrated, interpreted, and displayed.

Many observers thought it would be impossible for the biospherians to do all the work necessary to feed themselves, maintain the gigantic apparatus, and accomplish their research and development missions (on fifty hours or less a week). They could have twenty extra hours for communication and reports. Other demanding specifications: temperature had to be maintained between fifty and ninety-five degrees Fahrenheit and to average seventy. The machine shop had to repair anything inside without outside input. Most daring of all, in some ways, its tropical ecologies had to adapt to temperate zone variable sunlight. Therefore, its glass exterior had to cut down the sunlight as little as possible while being strong enough to resist hailstones, bullets, tornadoes, and so on.

One management task needed my careful attention for hours every day. Since I was the only one who had worked in so many of these separate fields and I had instituted our synergistic approach, that task was melding each of the seven biomic task groups and the technospheric task group into a creative unit, making sure that the results of all these creative lines did synergize. To achieve the first level synergy of each subsystem, every engineer, energy specialist, architect, technician, botanist, zoologist, mycologist, microbiologist, ecologist, economist, and manager had to work together. The objectively tough problem: tolerances. How much play could be tolerated in variation from the design once we agreed? There must always be some play.

Ecologically, Biosphere 2 contained seven biomes interacting to produce its world of life; dramaturgically, it contained seven dramas interacting to produce its world of action. Many top ecologists predicted it would be impossible to keep the biomes from mixing due to the migration of seeds and animals. Some thought it would all be taken over by some fungi. I didn't worry much about that because twenty years of work and observation had showed me how biomes defend their turf.

The first time I had been struck by the reality of biome integrity had been in Liberia consulting on a Swedish iron mine project. I was taken for a walk through the forest to where it abruptly ended. I looked in the bright sunshine and saw a continuous line of trees ending at a clump of savannah grass immediately at my feet, while ahead, right and left, the horizon stretched without a single visible tree. Obviously, the climate and soil didn't change abruptly at exactly that line. The different root, shade, and subsoil systems had met in a face off.

Dozens of people from different professions and cultures arrived to play different parts in our wildly different dramas and drastically different biomes. This new planetary theater concentrated in one location all our old theater and ecotechnic tours. The many varied, often apparently conflicting scenes on the site emerged as a unified dramatic action set in motion by humanity's conflicting intentions toward Biosphere 1 which were, therefore, reflected in Biosphere 2.

No one could escape the serious implications of this dichotomy. For example, if Biosphere 2 could produce clean potable water, pure air, waste recycled into flowers and food, record agricultural productivity (without chemical assistance) even though there were plant-eating insects inside, then why not do this in Biosphere 1, Earth?

My intentions: to create a decisive beginning of biospheric science and engineering; and to create a living evolving biosphere. My aim, my hope: for Biosphere 2 to become more than a set of operational algorithms; a transforming experience for those involved; for it to become a symbol, metaphor, legomenon, an inspiration to assist cultures, including those that would arise on moons and other planets and those existing now and in the future on Planet Earth, for biospheres and cultures to become co-operating partners to help manifest ever-finer tuned, more delightful complexities and beauties.

This task took my last gram of intelligence and energy even with two hundred of the best consultants on the planet and a magnificent team of colleagues and friends helping me in every way they knew how. Margaret must and would handle the executive side and Ed the chairmanship. Margaret was a genius who never lost touch with the doable; Ed was thoughtful and exceedingly thorough; I trusted them not to double cross me or the project. Biosphere 2 was also building itself in a sense; its historical necessity attracted the needed helpers because so many of the best and brightest saw that it showed new ways for Biosphere 1 and us humans to deal with our present crisis.

Ultimately, "The Biospheres themselves will give me the creative feedbacks that will finish the job if I stay true to them." This was my delusion, hallucination, illusion, dream, lucidity, poetry, mantram, my *idée fixe* at every level of my consciousness. When pressures got to me, irresistible forces set me walking in whatever bioregion of whatever biome my organism found itself, including the trial biomes we experimented with. Data and physiological regeneration poured in. My strength would return.

Like Antaeus who always found renewed strength touching earth, almost exhaustless energy flowed into me from contacting soils, grasses, trees, animals, breathing fresh air, drinking spring or rain water, eating its tasty foods while meditating on me in my turn feeding other life forms. I found myself trying to run with the coyotes at night, but they never let me get next to them, though they would circle close if they found me sitting down after a chase.

CHAPTER THIRTEEN

The Biospherians

In the last analysis, somebody has to do it.
SERGEANT JACKSON. U.S. SPECIAL FORCES

*We have found that as long as they eat together, morale
is being maintained.*
OVERHEARD AT A NASA CONFERENCE ON GROUP BEHAVIOR

*Ultimately, task acceptance determines whether a group action
leads to personal development and accomplishment or to the
eruption of subconscious forces of fight or flight, dependency-
kill the leader alternations, or pairing.*
FROM A W.R. BION GROUP BEHAVIOR EXPERIMENT AT
HARVARD UNIVERSITY

THE MOMENTUM OF BIOSPHERE 2'S PROGRESS and its inspiration to systems scientists, teachers, and the public attracted a quarter million visitors a year to the construction site. Plants and soils in terrestrial biomes and the mangroves, and coral reefs in the ocean biomes had settled into their new surroundings as the day of closure neared. Major television documentaries aired in Japan, Germany, Great Britain, and France. After checking out the site for several days, Phil Donahue, the talk show host, did a major presentation on Biosphere 2. The media began to look at what was happening there.

Biosphere 2 was paradise, hard work, yes, but paradise for me until that point. A private company, with no public funds, no association with huge corporations, but with many friends, great scientists, avant-garde artists, and skilled technicians. The public didn't even know the word biosphere, by and large. We often had to spell it out on the telephone when ordering something. For years, I had worked hard to stay as anonymous as possible.

IMAGE ON LEFT: *Biospherian crew, Mission One: 1991 - 1993. (Standing from left to right) Taber MacCallum, Sally Silverstone, Linda Leigh, Mark Van Thillo, Mark Nelson, (center) Roy Walford (front left to right) Jane Poynter, Abigail Alling.*

Abundant wheat crop in Biosphere 2 agricultural biome waiting for harvesting.

From the beginning, the selection and final training of the biospherians was a focus of attention. Of course, I did consider the possibility of simply hiring caretakers to go in and out daily; to run the world's most complex gardens; to eat breakfasts and dinners outside; and thus fairly easily do exemplary and non-controversial science. There would have been hardly any media scrutiny or time wasted trying to answer some wild statement. The science would have become another technical achievement and would not be considered a threat to reductionists picking at parts of Biosphere 1 such as the CO_2 level.

My aim was for the wilderness biomes to be operated "naturally," as much as possible like a wilderness area where entry is restricted to scientists and a few nature lovers. But it also called for the agriculture and human habitat biomes to be normative rather than imitations of conventional industrial agriculture and energy-guzzling, polluting cities. This meant that the biospherians or bionauts would have to master a demanding array of different tasks.

To produce a diet with the best ratio of carbohydrates, fats, and proteins as well as trace elements and vitamins, those who lived inside would have to plant, tend, harvest, process, cook, compost, and store the seeds for about one hundred fifty cultivars. They would have to do all this on about forty percent of their working time. They would have to do nearly all operations manually, though there would be excellent tools like seed thrashers, drying ovens and a top-flight irrigation system. The eight would do all this on little more than half an acre, or two-tenths of a hectare.

They would experiment with a new kind of agriculture, one that integrated ecology and physiology, advanced technics, a way of life, health, and sustainability. If they failed, it would shut the experiment down. However, if it succeeded it would become the first step toward a solution to living on Mars and an alternative to be considered in enhancing life in tropical regions of earth, introducing large scale local greenhouse farms close to city markets in the temperate regions.

Having to live in the human habitat instead of in conventional offices with immense technical support, they had to learn all the skills needed to operate the first paperless office with video, computer and telephone links around the world, a challenging feat in 1991 before the advent of the Internet. Most importantly, they could not be specialists concentrating more and more on less and less, ignoring or feeling guiltless about the wider consequences of their actions. Vernadsky had insisted that, ethically, scientists should consider themselves responsible for the consequences of their actions. They should not develop a toxic chemical or a destructive, polluting machine and then turn it loose on the biosphere for war, exploitation, or mischief.

Living inside, Biosphere 2's inhabitants, whether they wished to or not, would bear the consequences of their actions in a fast feedback cycle so their ethics (behavioral norms) could not avoid connecting with their epistemics (knowledge). Their atmospheric carbon dioxide cycle was approximately one thousand times faster than Earth's. One day equaled three years. Carelessness would quickly result in bad water, bad air, bad waste recycling odors and germs, bad food, wrecked communications, extra work because of sloppy maintenance, and dwindling biodiversity. Quick feedbacks meant causal relationships could be quickly discovered.

They would have to become as aware of their ecological and technical systems as their cultural system. In addition, their diet and physiology would be measured exactly so that they and any student of the data would see the consequences on their health, alertness, and vitality.

Because of the unique requirements for success as a biospherian, biospherian training engaged my thinking from day one. Fortunately, since first receiving the vision of an artificial biosphere, in 1969, I

Mission Control office inside the habitat was designed as a paperless office. This was back in 1991, well before the advent of the world wide web as we know it today. We had an on-site computer network and communicated with the crew via email, hand held radios, telephone, and direct communication at the window.

FROM TOP: *Sally Silverstone, Jane Poynter, Taber MacCallum*

had engaged in training myself and a few close colleagues in becoming biospherians living in Biosphere 1. Achieving this objective had caused me and my friends to travel at least once to each type of biome on the planet. We called these trips eco-expeditions and coupled them with theater tours so as to encounter in a deep way members of cultures that have adapted to these biomes.

Those of us who had become SBV board members had participated in all the Institute of Ecotechnic conferences that built so much of our intellectual capital. Two of the biospherians had worked with us since the early 1970s. Notably, Mark Nelson had played a key role in our whole set of operations since 1969, especially as chairman of the Institute of Ecotechnics. Mark was the only member of the SBV board of directors to become a member of the first crew to live inside Biosphere 2. He worked throughout Mission One assisting Margaret and me in the world of scientific politics. The brilliant, eloquent, witty Dr. Roy Walford and I had become the best of friends when he assisted TAP while we worked on the *Heraclitus*. Roy was a tough and welcome critic who made sure of the biospherians' health. His and Mark's papers are essential to understanding Biosphere 2. Without Mark and Roy, I'm not sure we could have accomplished the two-year mission.

The other biospherian candidates had worked on the *Heraclitus* at sea, at Quanbun Downs or Birdwood Downs and proved their resilience and cooperative knowhows under difficult and isolated field conditions. Mark Van Thillo had apprenticed in ecosystems at Synergia Ranch and ran the engine room on the *Heraclitus*. Jane Poynter had worked at Synergia Ranch, acted in the theater and took care of horses at the rugged Quanbun Downs project. Sally Silverstone had worked in the Las Casas rainforest project in agriculture and rain-forestry. Taber MacCallum had been a diver on the *Heraclitus*. Linda Leigh, a professional range ecologist, became a candidate in 1985, and Abigail Alling, a marine biologist and cetacean specialist, signed on in 1986. Each had special field and laboratory experiences that qualified them to start their training. Linda went on to get her doctorate in systems ecology under Howard Odum at the University of Florida; Jane and Taber started up a business consulting on space related projects; Alling, together with Silverstone and Van Thillo, manage the Planetary Coral Reef Foundation which specializes in coral reef studies using Phil Dustan's assessment method. All continued with biosphere related activities.

I decided that the most important element in their training would be to help build Biosphere 2, especially the areas in which they would be most highly engaged. Abigail Alling took charge of putting in Walter Adey's marsh and ocean biomes. Linda Leigh coordinated putting in the continental wilderness biomes. Mark Van Thillo was in charge of checking the equipment going inside. Mark Nelson organized scientific conferences and waste recycling. Sally Silverstone and Jane Poynter dealt with the crucial agricultural system. Taber Mac-Callum took on the analytical laboratory. Roy Walford designed the critically necessary medical laboratory and put together a superb medical support team – much depended on Roy. The seven biospherians of Mission Two had less vigorous training but did very well.

Margaret and I coordinated the training schedules with work schedules so that they reinforced each other. Eventually, they did a satirical play on problems that could come up called, *The Wrong Stuff*. That fundamental sense of humor certainly helped them.

Roy Walford kept Margaret and me, the board, and the Scientific Advisory Committee apprised of the health of the biospherians. Their weight charts provided as excellent a check on the progress of the agriculture as the carbon dioxide did on the state of the total system. Roy found that the Biosphere 2 diet produced a de-aging effect on the biospherians. Roy's results boosted Margaret's and my confidence

FROM LEFT: *Linda Leigh, Walter Adey and Abigail Alling doing field collections for the ocean and marsh biomes.*

ABOVE: *Mark Van Thillo working on equipment maintenance in the Biosphere 2 machine shop where he would have to repair anything that broke down during the two-year experiment.*

The crew meeting in the dining room to review the activities for the day.

in the whole operation. As long as good health continued and the biospherians ate together despite the usual strains arising in small groups, the system was on go. Roy also made interesting videos of biospherian life; several of the crew participated in "inter-biospheric art festivals" via videophone links where poetry, video, music and art were shared. Roy, who had been a wrestling champion in college, engaged Taber, who was nearly forty years younger, in a spectacular birthday wrestling match in a muddy rice paddy in between harvest and replanting.

Mark Nelson, an excellent horticulturist, took decisive charge of the crucial wastewater recycle system. He also served as the ever amiable communications officer. Today, Dr. Nelson speaks on and installs ecological Wastewater Gardens® systems all over the world. Co-author with me of *Space Biospheres* and my collaborator on many scientific papers, Mark's years of chairing Institute of Ecotechnics' international conferences gave him an easy way with the public, scientists and dignitaries. As an SBV director, he gave the board direct accounts of his take on what was happening inside the experiment.

I don't write about the biospherians so much in this book because it's about *me* and the biospheres, not me and the *biospherians*. I admired them all, individually and as a team. I tried to keep *me* out of it as much as possible when working with them; they had a hard enough job, and they did it with devotion, honor, and competence. They proved the most important point about humans living in a biosphere; they lived there for two years and left it a more beautiful place, maintaining good health and high spirits.

I'm happy they all found a way to continue their connection with biospherics. Their work in Mission One serves as an example to all of us and an incentive for humans to adopt biospheric principles. Margaret and I promised them all we would never discuss their personal lives. Of course, Roy and Mark were different from the others and special to me, project or no project. We went back a long way before Biosphere 2. We inhabited the same scientific and artistic milieus, loved satire and double entendres, were nuts about world history and were fellow philosophers. We each had a loner streak with a sometimes sardonic sense of humor directed against ourselves. Ultimately, we were not in any way, shape, or form believer types, not even in biospheres, not even in ourselves, and especially not in whatever our persona might be manifesting at any given time. Roy and I had knock-down drag out disputes about scientific matters (sophisticated reductionism vs. biospherics) but we wound up at our last two meetings (before he died in 2004) agreeing that all ways of doing science have validity at the right time and place. We could go at it hammer and tongs because Roy and I understood each other really well. We counted on each other as, Mark and I do at IE conferences where we balanced several different, even opposing scientific viewpoints.

Above all, Margaret and I had confidence that this group of diverse individuals, qualified in skills, health, and adaptability, were devoted to the biosphere and its well-being, that they revered the world of life, whether it be Biosphere 1 or 2. By 1991, the biospherians had

Roy Walford, above, and Mark Nelson. Roy completed Mission One in 1993 and returned to UCLA to work on his research and writings. Roy was stricken with ALS (Lou Gehrig's disease) and passed away in April 2004.

trained for at least five years at the most thorough level we could devise. Visiting astronauts often said our training was comparable to theirs. These eight (selected from fourteen qualified candidates) had the mindset to succeed in the Herculean task of lifting Biosphere 2 off and keeping it flourishing for those crucial two years. They heard the call and they did it.

To make my own intentions clear, in 1989 I told the board that I planned retirement from direct management in 1994 (at age sixty-five), after the start of Mission Two even if my perfect health continued. I could continue to participate in strategic planning if they desired. I had finished the second volume of a trilogy on my hero, Joe Madison, *Journey Around an Extraordinary Planet*, while at Biosphere 2. I wanted time to finish the third volume, *Liberated Space*, with Joe Madison arriving in the late Sixties in San Francisco, and do a book of poems, *Off the Road*, about those nowhereville kinds of situations that always liked me.

I still "wanted to become a writer." Exploring Biosphere 1 and designing Biosphere 2 had given me a wonderful life, copious and striking material to contemplate, but I wanted more than to enjoy it. I wanted to understand *me* and that material, and, besides, writing appeals to *I* as much as action appeals to *me*. Since I spent an average of two hours a day reading ever since I was eight, I wanted to read about my own life and circumstances by the one author who had authentic material.

In addition, I had a whole series of characters and plots for novels about the sensibilities of the seventies and eighties that later became a book of short stories, *My Many Kisses*. Several scientific papers about Biosphere 2 demanded to be written. Many biospheric discoveries needed a full explication. Creating Biosphere 2 made me ponder new notions of the ethnosphere – a technically-linked communion of peoples had to be developed. I could already see that my next major line of scientific work would be on aspects of the ethnosphere: the "Origins and Futures of Cultures."

From insights gained in designing Biosphere 2, I also had a paper on human taxonomy to work on. This paper arose from my having to design a biospheric model of Biosphere 2 that would analyze the human actions that had to be supplied and measured in terms of food, water, and waste interactions, as if humans were members of the animal kingdom.

But humans also have to have a special habitat that supplies the means for them to lead a full cultural life, with music, theater, a library, laboratories, atmosphere, privacy, and commons. This whole world, to be a true model, must manifest in beauty "the splendor of truth." In short, depending upon the observation being made, the humans in Biosphere 2 were either a group of big-brained animals or sophisticated members of a culture; their true humanity would consist in making their own integration, a third being arising out of their double being.

So I had to design two systems in this biosphere to accomplish two purposes – one to satisfy humans as physiological individuals (in subphylum supercraniata) and one to satisfy humans as culture creators (in kingdom *Symbolia*) – and then mesh the two designs into one biosphere. This made Biosphere 2 essentially a model of Biosphere 1. Margaret was an extraordinary co-creative agent in our detailed realization of this design. "Genius is in the details," she reminded my initial bubbling enthusiasm. Salutes to Margaret.

I proposed the "Biospheric Uncertainty Principle" to deal with the two worlds humanity lived in. Quantum mechanics discovered "uncertainty" as an inherent component of its science. Because a photon is so small compared to the impact of the observation of a photon, it could be seen as either a wave or a particle. In Biosphere 2, (and

Biosphere 1, of course) a human is so small compared to a biosphere-level observer, that a human assumes two simultaneous aspects, causing an uncertainty resolved only by which aspect the observer decides to examine. This unavoidable uncertainty about what a human is depends on whether the observer measures *meta*-bolic or *sym*-bolic activities. In one, humans are viewed as a part of a biosphere, a *me* organism; in the other, as part of the technosphere-ethnosphere, a set of intellectually-operated symbolic programs, an *I*.

In the case of this book, *Me and the Biospheres* endeavors to create an "objective correlative" to my metabolic sensations and emotions, my origin, rapport, and re-merging with life. I, who am writing about *me* in this book, however, is a symbolist who also points to my way of thinking about rapport with words, diagrams, and photographs. This may trigger the reader's own *me* to experientially investigate what are called biospheres and the reader's own *I* to experimentally test some of the algorithms. Meta-physics (beyond nature) endeavors to attain a worldview that has this "binocular" vision. So this book is also an expression of a philosophy.

The use of this principle of alternation of metabolism and metasymbolism played a decisive role in the invention of Biosphere 2. It resulted in a design that integrated the dual human components into one structure. One part of the design ensured taking care of all a biospherian's organic needs, the other allowed expression of a biospherian's cultural/symbolic/thinking needs.

In cultures that evolved without conscious design, as adaptations to the conditions of Biosphere 1, these two aspects appeared in the creation of two lines of prestige and power, such as shaman and chief, artist and scientist/technician, philosopher or clown and emperor, judge and president, unified by watching the inputs and outputs at the boundaries of the sacred region or homeland. Humanity cannot be truthfully classified as either solely a big brain primate species or

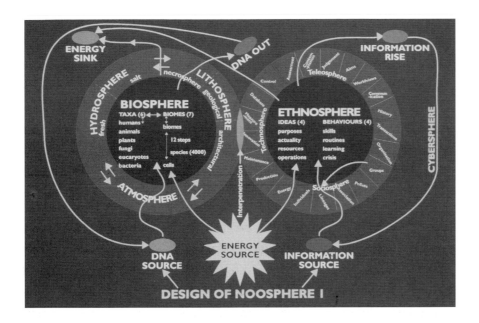

Design of Noosphere 1

a godlike manipulator of a series of algorithms. That becomes quite clear when one begins to study or design a biosphere that includes humans. The diagram above shows how the breakthrough design that "came down" first looked (actually second looked; the first look was scrawled on a coffee stained sheet of white paper).

This design galvanized me and revolutionized my thinking. It gave me all kinds of new ideas about theater (on stage, in my mind, in the midst of life) revolving around locating plots in three dimensions of inner time rather than in the four dimensions of space-time. It gave me new directions for projects: how to better realize the proven potential of the October Gallery, the *Heraclitus*, Las Casas rainforest, and Birdwood Downs' savannah systems; and how to approach the next step in biospheric science, designing an Earth to Moon and Mars prototype biosphere and a cybersphere for members of *Symbolia* to play with in electrons till it became comprehensive and quick enough to apply in Biosphere 1. And, intimations about how to write a book to take forward Vernadsky's concept of noösphere.

Best of all, the design gave me hope for abundant life and new projects after completing the twenty-five year effort to make and realize Biosphere 2. Whatever happened to me, I would do my duty: complete Mission One, establish biospherics, protect the interests of the shareholders in SBV and BD, and do what I could for the interests of co-workers that had pitched in for minimal compensation because they wanted to be part of the history. I felt beholden to all these comrades for giving me such a grand opportunity to do cosmic scale science and technics.

I owed the many people who believed in my integrity, that I would carry through to the end my promises about Biosphere 2. Strong feelings and occasional revolts surged in me regarding the personal costs associated with introducing new ideas and realities, with establishing biospherics, a systems science, in an exploitative, reductionist world – but the hope of succeeding always won the day.

But too many duties prevented me from continuing my daily long strolls, tranquilly soaking in the desert and letting my creative subconscious refill. Of course, short strolls continued, fifteen or twenty minutes nearly every day. Some days got so crowded that I ran the eighth of a mile between Margaret's office and mine to keep up my health and schedules. I carefully visualized returning the ship of Mission One to port on September 26, 1993, painstakingly measuring the results, and then setting sail on Mission Two before disembarking and being piloted to shore on my sixty-fifth birthday, May 6, 1994. This decision (based on my long practice of gaining energy from setting personal deadlines and cut-off points) gave me renewed vigor. My favorite health measurement was the rate of my daily hour's high-speed pruning of the trees that I had ecoscaped (sustainable landscapes) to enhance the fifteen-acre grounds surrounding Biosphere 2. By 1989, the growing complexity and momentum of the Biosphere 2 project thrust me into situations that forced me to drastically upgrade

John spent time nearly every day working on the grounds around the property, especially pruning the many fruit trees we planted. Mission Control is in the background.

myself or get out of the kitchen. A strange sentence that I don't pretend to understand fully came to me at this time: "The evolutionary secret protects itself and chooses agents by multi-level jumps to strange transfers of consciousness, agents who may be aware or unaware of what ends they serve and what battles they will find themselves engaged in." So I "girded my loins," as Fritz Brennecke, my old football coach at Colorado School of Mines, loved to advocate in his half-time talks. Late 1989 felt like my "half-time" at Biosphere 2 and at that point it looked like a tougher second half to me. To survive with honor, I would have to rev myself up to finish Mission One, testing how stable interrelated cycles developed in closed life systems and if humans could live as co-creators (and not as clever parasites) within a life-giving biosphere.

After Biosphere 2 was sealed, Margaret and I monitored operations by means of daily visual inspections. We did these separately to make sure we made independent observations. Each week, through the glass, we checked the state of our special emergency procedures, our

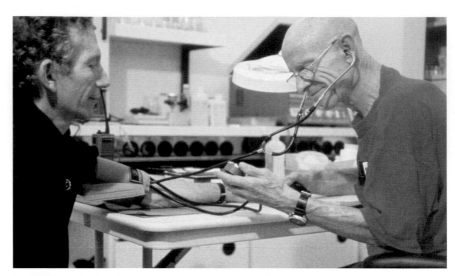

Dr. Walford had a rigorous program to monitor the health of the crew during Mission One and has published the medical results in numerous journals and books.

over one thousand automatic sensors, and the back-up laboratory. Roy Walford's well-equipped medical lab and top-notch skills meant that there would be close tracking of the biospherians' health. Margaret's team of medical consultants contracted to support Roy in any and all emergencies. My wonderful team of advisory scientists all contributed their services; only their expenses to visit the site were covered. The Odums, Gazenko, Runcorn, and Schultes, provided not only intellectual but moral strength that helped carry me through. Without Margaret backing Roy, and our access to Russian medical experience through Lydia Somova from Bios 3, things would have been almost impossible for me; my attention might well have reached a breaking point.

Roy never desisted from searching for and finding ingenious new ways to measure and monitor humans' physiological response to living in such a unique world. Biosphere 2 enabled exciting new findings to be made about humans in relation to this measured environment. Roy, one of the most disciplined men I have ever known, performed every duty to the hilt. His findings alone justified the experiment by proving that health can be maintained at a high level in a two-year closed biospheric system, and therefore, by extension, indefinitely: on Mars, in Antarctica, *and everywhere in Biosphere 1*. Roy's cutting-edge general practitioner medical office in Biosphere 2 could be put in every neighborhood and small town and provide the basis of a biospheric medical system in which all data would go to the planetary knowledge base. Roy published a series of brilliant papers spelling out these discoveries and possibilities.

None of the biospherians ever lost sight of the primary task that they had signed up and trained so hard for: that Mission One must be carried on for two full cycles in order to clinch validity of the data; without cyclic data, biospherics can't be done. The opposition of the reductionist scientists was so great that Mission One might be the one and only chance to get this data. All eight biospherians carried out their duties for the full two years without allowing their differences or outside pressures to divert them. They demonstrated the highest ethics and reverence for all life. Their behavior could well serve as a model for leaders to emulate in caring for their own bioregions and biomes.

Rembrandt and Mona Lisa

CHAPTER FOURTEEN

Mona Lisa is a portrait of Leonardo himself: the body of Greece,
the passion of Rome, the detachment of the mystic.
PARAPHRASE OF WALTER PATER, OVERHEARD AT THE LOUVRE

AS THE PRESSURES MOUNTED heading toward the closure of Biosphere 2, my inner sensations replicated almost exactly those I had experienced under orders by that clever disciplinarian (as he thought of himself), Corporal Salvatore Lacca, in the Korean War. I had been peacefully, even merrily, marching in step down the road at Fort Leonard Wood with my engineer platoon of Company B, when he started to switch my legs with a slender green willow branch – a painful, stinging process. An urge to turn and punch Salvatore, himself a trained pugilist, nearly overwhelmed me. However, the desire not to be punished for hitting "that no count blankety blank" and an even more overwhelming urge to prove myself, kept me swinging my legs in time to our collective chant: "I got a girl that lives on the hill," acting as if nothing unusual was happening that hot, gray, muggy day. Salvatore and I later became good army buddies, and I valued the painful leg switching as his rough-and-ready way of teaching me military discipline. Dealing with the media at Biosphere 2 reminded me of dealing with Salvatore: control yourself and keep your thinking on your objectives while being switched.

Recognition of Biosphere 2's unique value in producing new knowledge and experience grew rapidly at advanced corporate and cultural levels. Margaret spoke to top executives at IBM and Xerox and at the Air Force Academy (the Air Force had just realized the tremendous clean-up bills it faced at old airfields). I spoke at the Natural History Museum and at Manhattan's famed cultural venue, the 92nd Street

IMAGE ON LEFT: *Sunrise on September 26, 1991, Biosphere 2 closure ceremony. From left: Crow Medicine Man; Tibetan Geshe and Toltec Curandera, sing prayers at sunrise before the crew enters Biosphere 2 for the two-year experiment.*

YMHA (Young Men's Hebrew Association), in New York. Bill Dempster spoke at major national engineering meetings. These informed audiences gave enthusiastic responses to what we were doing at Biosphere 2. Two American journalists took up our offer to spend a day enclosed in the Biosphere 2 Test Module to familiarize themselves with what this new complex science could do. David Chandler from the *Boston Globe*, and Kevin Kelly, writing for *CoEvolutionary Quarterly* and *Wired* magazine, both visited Biosphere 2. Chandler's articles are an accurate historic read. Kevin Kelly's acute comments can be found in his classic work, *Out of Control*.

During the time just before closure, many visitors noticed me and said, "It must be marvelous to see your dreams coming true." Of course, that grand architecture, which allowed the elegant biomes of Biosphere 2 to flourish, emanating its own special atmosphere, kinetics, subtle light, and happy human sounds and insights, indeed marvelously realized my dream. In fact, materializing a project this complex and urgent wafted me beyond dream into lucidity. Biosphere 2 had become as much a late Rembrandt self-portrait, warts and all accentuating his wisdom, as it was Leonardo's *Mona Lisa*, which so harmoniously combines the body of Greece, the passion of Rome, and the gnosis of Nature.

As much Picasso's ravenous *Demoiselles d'Avignon*, as Matisse's fulfilled maidens in *Lux, Calme, et Volupté*. As much Artaud's Theater of Cruelty as Stanislavsky's radiant Moscow Arts Theater. As much Chartres complexity as Taj Mahal simplicity. As real as it was ideal. As much Heraclitus's eternal flux as Plato's eternal form. I felt all those dyads contributing their force to my life while sitting in the limestone sculpture on top of the artificial ridge's great amphitheater that framed the view of Biosphere 2. From that vista, I could soak in the scene without justifying or regretting anything.

My saving grace from stress was my twice-daily solitary walks around Biosphere 2 and my daily talks with Margaret, Bill, Marie and Deborah. These moments were reinforced by wonderful days with visiting masters like Richard Evans Schultes, William Burroughs, and Oleg Gazenko. Time passed swiftly as we explored fascinating details of and generalizations about plants, history, and destiny. The monthly talk given for the staff by our visiting scientific consultants, all outstanding thinkers whose friendship, knowledge, and critiques constituted an incomparable education in biospherics, were intellectual feasts. The real undying marvel, my biggest dream – that true pals could do wonderful things together and stay friends without being torn apart by ambition, greed, and fear – kept developing its story line.

Pallenville

CHAPTER FIFTEEN

I dream'd in a dream
I saw a city invincible to the attacks of
     the whole of the rest of the earth,
I dream'd that was the new city of Friends,
Nothing was greater there than the quality of robust love,
     it led the rest,
It was seen every hour in the actions of the men of that city,
     And in all their looks and words.

WALT WHITMAN

WE POST-TRIBAL, POST-HUNTER/GATHERER, post-modern humans live in *dystopia* (a lousy place). We have been stuck in a situation of patriarchal family, oligarchic property, and a military-based state with all its accompanying repression, wars, epidemics, devastation and famines. Ideological *utopias* (a nowhere place) pretend to solve "the problem" while ignoring the situation and have proven to be a succession of mirages. Perhaps *practopia* (a practical place), based on being as real and as simple as pals, can at last stick its nose in history's tent.

One afternoon, after napping for a few minutes by my favorite saguaro cactus, a half-mile brisk hike over the ridge from Biosphere 2, I regressed to the wide-eyed magical perception of a nine year old. I walked the ten blocks from home to school on the first day of that school year, in a new town, Okemah, named for a Creek Indian chieftain. My father had shown me the way to school the previous day. I never stepped on a crack in the sidewalk. Looking down, I saw small groups of two-inch-high humans making their way through a vast elephant grass savannah.

IMAGE ON LEFT: *Birdwood Downs sunset picnic with pals in Regeneration paddock.*

After surviving gripping scenes of hunger, thirst, exhaustion, and discovery involving these tiny humans (whose emotions and thoughts were bigger than mine), I arrived at the red brick school. The most interesting boy I'd ever seen in my life magically appeared. Though taller and bigger than me, I hoped and hoped he would be in fourth grade, too. He moved subtly, clothed in grace, power, and a white T-shirt. My heart stopped. I hid behind a tree to absorb his every action in the playground. I had to know as much about him as possible before we connected in what I already knew would be a meeting that would reveal my future. The other boys and the girls obviously regarded him as their chosen leader. I wished for nothing more in life but for him to become my best friend. I slipped out from behind the tree to risk the boldest move of my life. The distance between us miraculously vanished; we were staring at each other, dead silent. "I want to be your friend," I said. From that moment of infinite clarity till his too early death, John Tilden Palmer and I remained best friends.

John and I did a lifetime of adventures together in the next five years while I lived in Okemah. We practiced alternating Tom Sawyer and Huckleberry Finn personas. One day, at eleven years old, we stopped (hand-powered) mowing a half-block of grass to sip cold lemonade; we leaned back in wicker chairs and delicately placed the callused heels of our bare feet on his family's porch rail. Mowing lawns and raking leaves earned us our spending and saving money.

I also was writing and publishing (with a ditto machine) my two-sheet weekly newspaper, the *Chit-Chat*. It sold for only two cents, but as sales were good and you could go to "A" movies for twenty-five cents and "B" flicks (or Westerns) for ten cents, it covered my weekly movie money, plus an additional seventy-five cents.

As our relaxation deepened and that cumulous-cloud-punctuated summer day cleared our minds, John and I began designing our

John with his best friend, John Palmer, in Oklahoma, 1943.

yearned-for way of life. We sternly pledged to each other to do everything necessary to make it all come true. In this community, first and above all else, everyone was a pal. People could do anything they wanted to, think what they wanted to think, feel what they wanted to feel, as long as they stayed pals. That was the only rule needed, and the only rule made. We took our last names, Palmer and Allen, and came up with *Pallenville*.

Pallenville should do some good in the world. We gazed in mutual fascination at the piles of freshly cut grass that lay strewn over the surface of the half block to the side of the porch on which we sat like caliphs from our abridged version of Burton's *The Thousand and One Nights*. All Nature was our empire. A flash hit us. We turned to each other united in a single vision, one life-changing glance. Fantasies about telepathy have never imagined such perfect communication. If we could invent a way for humans to eat grass, all food problems in the world would be solved!

We gulped down our lemonade, ran down the steps of the porch, and selected the freshest, tenderest moist leaves of grass. First, we tried long-term chewing, just to make sure the obvious had not been overlooked. After three minutes of determined chewing, the mouthful of grass remained a hard, undissolved lump. Then we ran to the kitchen and inveigled the use of John's mother's pressure cooker. We examined a thick recipe book and found the answer. Spinach soufflé could be the historic forerunner of grass soufflé! Spinach soufflé could be done at Earth's atmospheric pressure, but grass needs to have the pressure increased to transform it into food for humans.

A pressure cooker should certainly be able to do all that a cow's stomach could do! We turned the pressure cooker up to "high" and cooked the grass until the pot whistled. We took it out and shared it equally, half on each plate. The cheese was delicious. The grass seemed definitely softer, and we chewed on expectantly. However, after endless grinding and the application of much saliva, the grass remained indigestible. Our appreciation of cattle and horses grew to admiration. Pallenville turned to other projects!

Palmer and I could only go so far together because eventually there arose the question of girls. In those days, girls couldn't be pals. After a few efforts at double dating, girls meant leaving your pals and going off with one of them. (After 1967, girls could also be pals.) But then, being pals is as far from being clone-like huddlers as you can imagine. Like, dislike – all that's noise from a lower level. Pals *are*, and what pals do emanates from who they are. The different things they do, feel, and think relates to being pals first and last. That's what John and I thought. I still think that.

Dreams are notoriously fraught with transience but dreams of pals endure. In his masterpiece, *Life is a Dream*, Calderon wrote: "What is life? A frenzy ... an illusion ... agony, and a dream. And the dreams themselves are dreams." But dreams, sometimes showing us a glimpse of Alph, the sacred river, sometimes transform into works of art where their insights purge you of the pity and terrors distorting your actions. Our dream of pals at Synergia Ranch had transformed, I felt, into the achievement of Biosphere 2. I often saw John Palmer in my mind, his broad shoulders and loosely swinging arms, his smiling view of my attempt to create our vision of Pallenville, his laughing, healing scorn at its obvious defects.

Unlike our early (and simpler) aspirations for Pallenville, the science and adventure of Biosphere 2 had to mesh with rational calculation and pragmatic action. Our engineering and architecture aligned a dizzying range of forms and functions with canons of beauty. The abundantly diverse and increasing biomass of the world inside provided delight with its variegated self-purifying waters, fields of high-grade grains and vegetables, bountiful banana groves, the waterfalled and mysterious rainforest, lively coral reef, stealthy mangrove marsh, perfumed desert, and rambunctious savannah. Invisibly, all these biomes and bioregions synergizing with light and microbes created a healthy and sweet atmosphere. On full moon nights your outstretched hands could touch … magic. But besides truth and beauty, happiness requires good – and good requires actions to be ends in themselves. At times, I felt conflict: Was I doing Biosphere 2 as an end in itself or to gain some external end, some personally desired result? Often thoughts floated through my mind on the order of: *I could use Biosphere 2 to…. something or other.* But my judgment prevailed: it was a good in itself; good to have done it, good that people saw it and good no matter what happened to my personal wishes.

As part of our preparation for Biosphere 2, Margaret and I had contemplated the fate of Shah Jahan, imprisoned at the Red Fort in Agra, taunted with a distant view of his masterpiece, the Taj Mahal. That very possible outcome for us from Biosphere 2 hadn't stopped us. My dream included leaving it behind after we finished Mission One. One of my favorite toasts is, "Finish one thing and on to the next." A crafty Bedouin in Jordon had once pointed out to me, over our cup of tea that he had managed to brew over three twigs, that a desert rat always had two holes leading to his temporary refuge, one to enter, and one to exit.

In my expeditions to far off and far out biomes and civilizations, I often wandered idly around paradisiacal and spacious gardens built to the specifications of somebody like myself. Some of those wanderings were incorporated into Biosphere 2 and its site. For example, a mulberry lane wound its way around the lower base of the great earthen sculptured wall that arched around the back of Biosphere 2, as if it were a natural ridge. That sculptured wall mountain connected the two original hills, one to the west and one to the north, to make a vast amphitheater from which to enjoy different views of the biosphere and the Santa Catalina Mountains. Along the top, I put a viewing lane with alternating mesquite and thorn scrub trees, spaced so that if you stopped between the trees, you could snap an inner frame with your attention or an outer frame with a camera, then mount the frames into an overall "film" of Biosphere 2.

I modeled the lower lane on the great mulberry high roads to which Anil Thakkar had taken me. Shade trees with their silkworms adorned royal roads in small kingdoms around the Rann of Kutch, where Krishna had been born. Poetic or tired wanderers could sit in cool shade on days hotter than the hottest Sonoran August. On occasion, while sitting on that very lane finishing my book of poems called *Wild*, I could "see" Biosphere 2 as if it were in a three-dimensional Antoine Watteau painting. The "mountains of Cythera" (the Santa Catalinas) beckoned me from beyond the canyon gorge that separated them and the site where we were embarked on our romantic voyage.

I had deliberately planned the scraper-sculpted wall mountain to fall so steeply toward Biosphere 2 that no one would run or slide down it, except on a dare. I planted cactus to deter the foolhardy. Toward the west end, just before this top lane began its descent to join the lower mulberry lane, you could look through our sculpture of great limestone rocks towards the Biosphere 2 tower, a peaceful, meditative place for me and many others. Margaret brought my design off with the uncanny precision of a transcendent artist. The feline drivers of our earth haulers, once they got the hang of it, formed the dangerous slopes with polish and power. This sculptured ridge reminded me of the ancient Amerindian mounds that snake around the Mississippi Valley, on one of which I had done my only archaeological dig. On the ridge, one could contemplate the gleaming tower and structure of Biosphere 2 and, across the canyon, the stark Santa Catalinas that put wilderness between us and Tucson. To the other side, one could gaze into the haunting northern end of the Sonoran Desert, terminating in rugged mountains full of abandoned copper mines and memories of old Apache wars. The ridge trail dropped gradually down to the research and development area on one side, and descended on the other to the waterfall evaporative water towers where you could listen to the splashing water that cooled Biosphere 2 in the hottest of summers.

Johnny Dolphin, left to himself without any daily duties, could probably have eked out some polished stanzas during his remaining years while gazing from the top lane, perhaps even developed a short story or two. My not hanging around stemmed from loyalty to my pals and from understanding that wandering in such a worldview, without real work to do, would quickly become preciosity. I hadn't spent twenty years of my life working to bring about Biosphere 2 just to get buried there. Nor had I spent over fifty years of my life ceaselessly working, traveling, and studying since Pallenville to become a phony. Palmer would have knocked that crap out of me, real quick!

Biosphere 2 should stand apart from me and do its own work as a symbol, a metaphor, and I must stand apart from it to continue my work. Biosphere 2's existence, even if it diminished to a tourist stop or transubstantiated into a set of photographs, drawings, scientific papers, and memoirs, highlighted the flaw of the New World Order imposed by a hegemonic global market honoring money as God. It showed all who could see that our stupendous Biosphere 1 system was a unity that included humans; neither it nor we humans can flourish if divided into pieces for financial or scientific exploitation.

Diploma factories churn out reductionist specialists who divide total systems into pieces. Hired by Global-Tech corporations and nation-states, they engage in dividing up and devastating Biosphere 1 under a perverted jihad called "development." It rolls over humans everywhere. Its techniques feature good rhetoric and bad faith. They produce dangerous chemicals, automobiles, superhighways, processed foods, devastated forests, poisoned waters, toxic air, anthrax, dirty bombs, invasion of personal lives, and so on, ad infinitum. Ethics and aesthetics are a drag to those generating super-profits and "prestige" by inventing new scourges. "Accredited" replaces ethical, "respectable" replaces aesthetics. Humans pay the price in sickness and in anomie, loss of meaning.

I suddenly saw that acting out my Mystery Epic, my drama of Biosphere 2, would result in my being condemned to seek and never find, until I realized (again!) that it's making the all-out voyage itself – to Ithaca, to Colchis for the Golden Fleece, to the island of immortality, or to Biosphere 2 – that gives entry into the transcendent. Reaching a goal means starting towards another.

My deepest hope in 1978, when we formed our consortium of Decisions Team, Biospheric Design, and the Institute of Ecotechnics, was for the dozen or so of us who worked together to become pals and to stay pals even if some of our projects turned out to be grass soufflés! We had all been through adventures enough to merit that elite appellation. Indeed, a cursory account of all our adventures would fill a book far bigger than this one. I envisioned us creating a new model of Pallenville; a true Caravan of Dreams. A dozen others who had passed through various tests of being pals were moving toward the same point of free association and disciplined action. Together, we had made practical life-enhancing projects where humans lived in harmony with nature.

As early as 1969, I thought these projects would somehow culminate in making Biosphere 2 a working model of Earth's biosphere. The model would uncover the laws and patterns of our biosphere's behavior, so we obstreperous humans could exist in harmony with Biosphere 1. Such discoveries would enable humans to prosper on Earth, Mars, the Moon, and perhaps elsewhere in the solar system. This would be the next Jeffersonian frontier and ecstatic Rimbaud-Whitman poem.

Pals play as hard as enemies to increase one another's awareness and strength. Palmer and I, for example, would wrestle until one could clamp his legs around the other's waist and tighten the clamp until the other's struggles ceased, from lack of blood and air. We challenged each other to create a sandlot football team of six, and we would play without pads. He downplayed this by calling his team "The Creampuffs;" I called mine "The Trojans." Games were played without quarter, but we obeyed unbreakable rules: no clipping, slugging, or holding. At pal level, it's all good clean fun because pals never do permanent damage; they don't clip and hold. They make everything right after they win; they smile when it's their turn to gasp in surprise. Nothing namby-pamby about pals.

In the spring of 1992, as the sun rose in the sky and the light lengthened in the northern hemisphere of Biosphere 1 and over Biosphere 2, I knew it had worked! Our agricultural output rose, the analytics lab poured out data, and the gloom and doom predictions that the experiment would break down after three months had disappeared. I lightheartedly and light-footedly broke into spontaneous short runs for the first time in months.

Biosphere 2, analogous to a physics cyclotron, was a complex apparatus designed to measure numerous interacting variables, not subatomic particles. Relationships weigh less than alpha and beta particles, less than quarks. But they too are charmed and strange. And Biosphere 2 was "unearthing" all kinds of relationships!

Gerald Soffen invited me to NASA's Goddard Space Center in Maryland to talk to his graduate students. I dined at his homey apartment with his charming Japanese wife, loved his wit, his stories about the Viking Missions, and poring over work-in-progress videos from the Hubble Telescope. Gerry had introduced us to Phil Dustan, who became essential to the coral reef ocean biome, and to John Corliss, who worked at Goddard and who would be my new director of research. Gerry took me to the Smithsonian, which I have loved ever since spending Saturday afternoons there while taking advanced basic training at Fort Belvoir. I marveled at things like throwing a Bailey bridge over a river, or cleverly stopping tanks from killing you. In my spare time ("hurry up and wait" was the army's mantra), I read the Smythe Report on the atomic energy project. It turned me on to large-scale science-engineering achievements. A nitpicking lieutenant didn't want me to read anything – "This is the Army" – but I bargained with him and got permission to read engineering and technical literature in my shop. That lieutenant let me change my life, which goes to show, you never know! That book led me to Colorado School of Mines, with the help of the GI Bill.

At the Air and Space Museum, Gerry showed me fine points of the Viking spacecraft, hung high up on display. He had commanded its Martian scientific program. When he first examined Biosphere 2, he hailed the project as "producing the same excitement as the Apollo Program." As co-editor of our Biosphere 2 NASA workshop publication, he had reviewed my "History of Biospherics" paper, so he recognized the full story.

He took his wife and me to see a play on Melanie Klein, whose groundbreaking work on group behavior I studied at Harvard. Gerry said, "I took you to this play because you and my wife like theater. I would have left after the first act except you two were so interested. I prefer my facts taken straight." Actually, I knew Gerry loved theater. He loved the drama of the space age and of going to Mars. That's how he fell in love with Biosphere 2; he saw it as a necessary part of the drama of going to Mars. We became bosom buddies for a while. Gerry just had the idea that drama only took place on theater stages.

I always maintain to actors that theater exists in three different localities: the mind, the stage, and in life. The same aim prevails in all three arenas, to reveal the inner life of humanity through the characters. However, rules affecting overt behavior vary greatly in these arenas, and so dramas can look quite different, depending from which of these locations they are viewed.

PART IV

A Thing of Beauty is a Joy Forever

CHAPTER SIXTEEN

Mission One Reveals a New World

THE TIGER OF BIOSPHERE TWO
In a hundred and eighty tons of air
Carbon dioxide prowls;
Lashing a tail on a stormy night
It picks up a hundred kilograms;
During the lazy hot day following the cold front
It exhales the same back into the green surround;
King of its three-acre domain,
Striped with daily ups and downs.
JOHNNY DOLPHIN

IN FEBRUARY 1993, WE HELD A MEETING of the Scientific Advisory Commitee to review Biosphere 2 science and examine and appraise our work. The science advisors consisted of: Eugene Odum, Oleg Gazenko, Keith Runcorn, Ghillean Prance, Harold Morowitz, Steven O'Brien, Gerry Soffen, James Arnold, and Tom Lovejoy. The engineering advisor was Robert Walsh of Allegheny Ludlum.

We were more than ready for the meeting. Bill Dempster and Roy Walford had presentations on engineering and human health. Margaret and I reviewed these so we figured we were well prepared, yet ready for anything unexpected. Biosphere 2 had revealed a science: biospherics. And the engineering had worked beautifully!

However, the night before the meeting, I dreamt that I was fleeing over the cactus, like Geronimo escaping from the fort where they were planning to hang him for fighting for his people's place in the

IMAGE ON LEFT: *A vision realized.*

biosphere. With poetic notions inspired by that vision, I got up at about two in the morning and ran barefoot across the back acres to see if this would induce such a state of organic realism that, like Geronimo, my feet would not be impaled by cactus. It got really interesting trying that trick at night; my feet never got bloody, even in the dark of a new moon and starlight pouring down through that clear and cloudless sky.

I have loved deserts since I first worked in one. It happened to be this same Sonoran Desert, back in 1949, where I picked cotton and fruit. My desert compadre, Tony Burgess, and I had been thick pals since making the desert biome for the Caravan of Dreams roof. We had examined a number of plants and ecologies to work out that exhibition of five different deserts. Tony and I discussed which desert biome concept to choose for Biosphere 2's biome. Tony was totally committed to qualitative exactitude and made sure that all the right plants and soil were put in place.

Dr. Walford made the crucial report everyone wished to hear. The conclusion: the health of the biospherians was excellent. Everyone was in fine shape to finish Mission One. Margaret asked if any biospherian wished for any reason to come out (with position and pay guaranteed to continue). No one did.

Margaret had made sure that our medical monitoring system was thorough; the system was a key part of the R&D. As CEO, she was determined that no one should come to harm on this project. Roy Walford's data provided irrefutable evidence of the biospherians' healthy condition; a team of doctors, including cheerful Don Paglia, from the UCLA School of Medicine supported his efforts. Dr. Gazenko, the world expert who had examined thousands of personnel and health reports working in extreme environments like space, high mountain closed systems and underwater, said he was quite happy with the data and with his personal examination of the appearance of the biospherians.

Dr. Walford in the Biosphere 2 medical office.

Margaret and I made sure that Roy's physiological and medical work be independent of other Biosphere 2 research. He reported only to her, the CEO, with copies to me. Systems ecology must include a focus on human physiology in biospherics. Roy, as most outstanding scientists who had run large laboratories for years would have done, thought he should direct all the scientific work inside Biosphere 2. Roy made his strong feelings about my "mistake" very clear to me. "I was enraged," he told me with the faintest of smiles in 2002 as I partook of a jigger of his aged Scotch whiskey in his Venice, California studio. I said, "I love you, Roy." He said, "I love you, John." Such is the sweetness of moments when all transient conflicts have fallen away.

After the all-day meeting reviewing the data, asking penetrating, challenging questions and taking comments with all the biospherians hooked up by video, our nine advisors accepted our data, reports, and analyses of Biosphere 2's state. Biosphere 2 had passed a crucial test, confirming that it revealed a new world, with its own atmospheric cycles, good human health and creativity, flourishing biomes, complete waste recycling, its synergy of advanced technosphere with a prospering biosphere and with open communications with its parent world, Biosphere 1. I felt that this free information exchange between the two biospheres was the most historic achievement. It foreshadowed biospheric co-evolution throughout habitable parts of the solar system and over the Planet Earth's "extreme" environments like Antarctica or Megalopolis.

There still remained much to do before Mission One ended, but Biosphere 2 had passed its thorough checkout. We could proceed with confidence. For me, this review was the climax. Now I began to prepare the Transition Mission so that all this irreplaceable data could be checked by scientists who could then themselves work inside Biosphere 2 once the eight biospherians had come out. I could move on to my next project, the understanding of the ethnosphere and beginning the creation of a cybersphere, a comprehensive feedback system of biospheric and technospheric data. Then peoples and cultures of Planet Earth would get necessary real time data; they could make a synergy between their world of life and their world of things and enhance their world of values. My next dream began to materialize, v-e-r-y s-l-o-w-l-y.

Before the Scientific Advisory Committee adjourned, Gazenko called me aside to say he must give me some urgent advice. He said that in dealing with high-stakes, high-visibility, historic kinds of experiments involving humans, one had, despite all the pressures, to maintain calm and perspective, and possess a deep knowledge of human nature. He had been attacked many times. He was on the international committee that set standards for such experiments and knew exactly what happened with medical experiments, as well as with cosmonauts and Antarctic explorers. Gazenko told us that SBV had not created enough documents, despite the many we possessed. He told us we must record the smallest moves in the progress of such an important and complex project – a project which encompassed not only deep human aspirations but also touched the raw nerves of the established powers in science, politics, the economy, and even the artistic world.

"You must create a Lego-land documentary trail that can be followed in detail by everyone interested. And you must keep secure copies. Otherwise, you will always be open to accusations by ambitious men who have few scruples to gain their ends." Fortunately, Roy had the medical documents meticulously organized and Abigail and Mark organized the ecological data. Of course, even with the most scrupulous documentation any project that breaks new ground will be attacked, that's the nature of any status quo, but one can also fight back and win. Margaret, Roy, and I put Gazenko's advice into immediate effect; the effort paid off big time and showed in later scientific papers, conferences, and projects.

CHAPTER SEVENTEEN

Murphy's Law and Finagle's Factor

MURPHY'S LAW: *If nothing can go wrong, something will.*
If one thing goes wrong, everything will go wrong.
The only way out is Finagle's factor.

NOW THAT BIOSPHERE 2 WAS IN FULL OPERATION, a number of astronauts from the Apollo moon landing program, like Buzz Aldrin, Joseph Allen, and Pete Conrad, dropped by to soak up the scene. Hundreds of outstanding people came from around the world to visit the project and share their expertise. Side by side with total system and biome behavior studies, I began specialist experiments on ecological subsystems inside with principal investigators like Howard Odum who obtained their own grants. Margaret and I began scheduling some of these experimenters into later stages of Mission Two, the way NASA did on space flights.

The last year of the closure experiment, two hundred and fifty thousand visitors enthusiastically took advantage of our real-time science tours. Deborah Snyder ran the education and publication programs: bookstores carried our line of educational books from the Biosphere Press. A satellite uplink program had been put on to an audience of a million schoolchildren in North America and had received a wonderful response. Young students started to build models of mini-biospheres in their science classrooms. Mail bags holding thousands of letters came to us from students with messages to the biospherians and questions about what life was like in their small world.

IMAGE ON LEFT: *First year anniversary celebration. The experiment was already setting world records in human life support research and we expanded educational outreach. One of the most refreshing aspects was seeing children respond to Biosphere 2 with genuine excitement.*

I designed Biosphere 2 on a sustainable, net-return-for-everyone, multi-vector model of economics. It seemed to me that the best thing would be for corporations and governments to synergize an expanded definition of land (the biosphere as well as property), labor, capital, and inventor/entrepreneurs rather than relying on the old formula of land (property only), capital, and labor.

Of course, individual fortunes and even short-lived empires can be made by buccaneering tactics. But a culture's people often fight for their land rather than see it raped by a foreigner's desire for profits. Simply dropping return expectations to six percent plus inflation would ease the pressure on everyone (even on the investing class because of fewer business failures) and on the biosphere. Inventor/ entrepreneur types should relocate to more biospherically rewarding corporations or countries where century-long rewards would be better.

This is something anyone would do if they took to heart Buckminster Fuller's definition of wealthy as opposed to *rich*: "Real wealth is the power of metabolic and metaphysical regeneration." That insight had put my life on the right path. Riches have in themselves neither regenerative power and, taken as ends in themselves, soon lead to degeneration of both metabolism and metaphysics.

In my play *Marouf the Cobbler*, the hero, Marouf, finds a secret: always keep a caravan coming his way, if only in his creative imagination. I first saw my "caravan of dreams" moving toward me from a great distance, over a vast desert. In my imagination I heard the tinkle of bells at our Caravan of Dreams in Fort Worth. Now I saw the camels coming over the ridge, as I watched the daily dramas of our discoveries about the *Real* World Order in the biosphere. In fact, I could already feel this caravan unloading gifts that would integrate my life with an assurance of metabolic and metaphysical regeneration. With the arrival of this caravan of realities, I could make deals with whatever, whomever, showed up in the bazaar of ideas.

I remember the start of wishing for a Caravan of Dreams, in 1959. I was head of Union Carbide Nuclear's two-stage, thousand-ton-per-day, crushing and leaching "B" plant at Uravan, Colorado. It produced uranium and vanadium concentrate. The ore was often hauled by mules that had been loaded by Navajo miners. The Indians had mined this area for uranium oxide to use for yellow paint long before the discoveries of Madame Curie. Later, I discovered that many of these miners died prematurely from radium exposure. Just a mile down from my plant, an old wooden chute snaked back and forth, slowly angling down the red sandstone cliff. Decades before the radium boom, Chinese prospectors had built the chute with hand tools to sluice for gold.

Western Colorado and I got along real fine. I took long hikes cautiously working my way with my buddy, Bart, along narrow foot trails on sheer canyon walls. Bart could spot an old arrowhead nearly buried in the clay, but my silent desperation saw no way out of a lifetime at the margins of meaning. I wrote some poems meditating past dreams of Indians and frontiersmen. A first camel arrived surprising me with some genuine lapis lazuli, clueing me that caravans did exist.

The president of the Colorado School of Mines, Vanderwilt, and the president of Harvard, Pusey, called me out of the blue to offer a two-year scholarship to Harvard Business School. It didn't take long to say yes, even though I would lose my salary for a couple of years. It meant having to do part-time work at the school to make up the

John's Harvard Business School photo.

difference (I ran the gymnasium clean-up crew). First, I could see how Union Carbide (and industry in general) regarded its employees as expendable at any level. Second, I really wanted to meet Radcliffe girls, famous for being the smartest and most independent girls in the world, at least in 1960. I had long realized that intelligent women waking up to reality and moving into the thick of the action represented the best, if not the only, hope for America and the world, and therefore for me personally. And a lovely Radcliffe girl did civilize me.

The CSM and the HBS had started this scholarship, the two presidents informed me, somewhat tongue in cheek, because "the Miners should become more refined, and the Harvards should become more rugged." They wanted the first recipient to succeed, and they thought I could.

But my biggest reason for saying yes was that after the Army Corps of Engineers, Colorado Mines, Allegheny Ludlum, and Union Carbide, I finally saw that Spengler and American technocrats (including my uncle Roy, a well-known athlete) had gotten it wrong. Engineers did not and could not run the culture of the West or any other culture. Yes, they were necessary for the performance of technics-based systems, but so were proletarians and soldiers and college professors. Engineers were treated simply as precision tools, powerless to shape society's forms. Now I wanted to check out finance. Perhaps this New Class did represent something creative in western culture; perhaps an innovator could find some leverage from which something new could be lifted into place. I had not yet realized that all forms of biosphere-exploiting cultures were being driven to a dead end by subconscious forces, imprinted by trying to maximize profits. Forests and soils and ways of life, humanity's unique heritage, the biosphere, were withering away, meaninglessly destroyed. Murphy's Law.

Initially, spellbound by my grandfather's skills and wisdoms, as well as by living frontier legends like Pawnee Bill, I had thought that ultimate sway in the American West's decisions was held by enlightened Jeffersonian farmer and rancher sages. Then, seeing firsthand that class of America destroyed by the Depression, and corporations taking over foreclosed farmlands and ranches with absentee owners who knew nothing of the land, recklessly spraying chemicals, I, at eighteen turned to romantic and avant-garde poets ("the unacknowledged legislators") with their visionary sensibilities.

At twenty-one, after hitchhiking to Greenwich Village in New York to see the Bohemian corrals of hopeful artists, I searched out philosophers but I found pragmatic operationalism only needed the status quo. Then I hung out with revolutionaries but found they produced gulags of bodies and ideas, and so I explored technocratic engineers only to find that their chief occupation was exploiting the biosphere more efficiently and profitably for the rich.

From these firsthand experiences and long conversations with real scholars of the Book of Life and Death, I realized by age thirty-one, roaming old Indian trails on weekends, that *none* of these forces could lead to creative change. They could only acquiesce or worsen the situation. In 1960, I had yet to see that coordinating notions of synergy, complex dynamics, and creative groups could end the tragic conflict between humans and their world. (At least it would end my tragic conflict with it.) Since masters of finance and control looked like the next would-be leaders, I needed to check it out to see if they held genuine power, that is, the ability and knowhow to initiate the new and better. Maybe I could find out the truth about them at Harvard, on the Charles River, and then move into that Arabian Nights on the Hudson, in Manhattan.

Years later, at Biosphere 2, it all began to come together. I began a new play with Kathelin, *Cyberspace Institute*. Murphy rules Earth's roost with his law that if nothing can go wrong, something will, and if one thing goes wrong, everything will go wrong. Finagle arrives on the scene to oppose Murphy with his famous factor, which allows his intuition to deal with Murphy's entropic rationalities; the ardent but battered heroes and heroines in this War of Des-Troy achieve a precarious balance in a biosphere beset with a menacing archetype still on the loose.

Biosphere 2's second winter went even better than the first, in spite of the big El Niño clouds coming in and again lowering the light energy. From the first year's experience, the biospherians had organically understood the cause-and-effect relationship between applied skills, how much they had to eat and how they needed to use Finagle's Factor. They could now compete with a Chinese or French peasant farmer. In other words, they engaged their senses and emotions as well as their intellects in digging, hoeing, harrowing, seeding, tending, harvesting, and processing. I had tremendous new hopes arising from this phenomenon.

Julian Sprung, our new coral reef consultant, helped us eliminate the old-fashioned algae scrubbers with a simple new device (protein skimmers) that took much less time to remove excess nutrients from the ocean's waters. Contrary to many expert forecasts, the carbon dioxide cycle month by month repeated the first year's cycle, allowing for the changes in the amount of daily sunlight due to different amounts of cloud cover. Biosphere 2 *had* self-organized a stable atmosphere! Water recycled, waste recycled, air recycled – all world records.

Experience recycled. The humans grew more food and ate better than in the first year and started regaining weight. My prediction (solidly based on biospheric systems science, Shepelev's and Gitelson's experiments, and our own Biosphere 2 Test Module results), that a comprehensively-designed complex life system would self-organize with sustainable stability, had proved true. Our experiment had been carried out in full view of the public, the media, and our scientific and technical colleagues and critics. All of this exposure, as we had planned, made us very careful indeed – a carefulness that helped us to avoid many mistakes (of course, many still occurred) and a couple of blunders.

With Mission One approaching completion, marking the beginning of the end of capital and investment by Decisions Investment, Decisions Team, and certain sponsors, the time had come for a close accounting of just how SBV could make, and how fast it could make, the transition to an independent enterprise. The Visitor's Center was steadily increasing in popularity and Margaret had started a promising publications industry under Deborah Snyder's strong leadership. I was developing a couple of inventions, but we had to get down to the details where Finagle's syntropy, as well as Murphy's entropy, loves to reside.

Robust, detailed, comprehensive, long-term data is the vital missing link in understanding biospheric processes. The value of this data to humanity is incalculable. Observing a system like Biosphere 2 in both the details and in the broad patterns of its functioning would allow scientists, for the first time, to evaluate quantitatively evolution and history working together on soils, atmosphere, hydrosphere, and biomes. It takes decades for some ecological cycles to display their variability. Biosphere 2 was designed for a hundred years. Long-term efforts, if continued on these lines, would produce a huge jump in fundamental scientific understanding. But whether Biosphere 2 would continue playing a role in such research would turn out soon be in the hands of others who would change the use of the facility dramatically.

The approaching end of Mission One and preparing to retire from the demanding project sent me off on long solitary walks. I grokked the stream-bed below with its challenging rocky narrows, or wandered out through the cactus getting out of sight of anything man-made. I found inviting shady bends in a sandy arroyo where I could watch my associations and dreams float by. I would turn sixty-five in a year and was more than ready to retire from Biosphere 2's daily multi-vector-leveled action. It had been wondrous and necessary, but always a battle. I didn't want to become a corpse, even with decorations on my chest. I had places to go, people to see, things to do.

I felt all my subtle energies activating again in every small muscle, just as when Sergeant Jackson challenged me to see if I could keep up on a twelve-hour run through the Vietnamese jungle with his youngest Special Forces sergeant. Otherwise, even with my MACV (Military Assistance Command Vietnam) stringer correspondent pass, he wouldn't let me go on patrol with him. It was how I had felt when the *Heraclitus* lost its anchor in a gale and was blown ever nearer jagged Akumal reefs.

I amused myself playing versions of the old game that Marie and I used to play at Project Concern, on the River of Dreams at Dam Pao, in Vietnam: how to escape through canyon and over cliff if the Viet Cong attacked. My walks through the surrounding Sonoran starkness took on ever more refined perceptions. How did the coyote get away from the armed hunter, the rabbit from the coyote? How did the starling escape the hawk? How did the snake get away from the hunter? How could I get away from Biosphere 2?

Occasionally, I plunged down ravines on my trail-wise gelding behind Clare St. Arnaud, my Ironman, horse-shoeing, tribal leader, Lakota brother, supple on his superb stallion, and I learned new wilderness tricks: how not to let a horseman get away from you. One of Clare's best-loved teaching jokes was to turn abruptly at right angles off a small trail and plunge straight down the steep slope of the arroyo while dodging cacti, mesquite, and sunning rattlesnakes.

My frontier phenotype and genotype began to actualize its potentials in ways that amazed me. I had always taken great pleasure in wandering wilderness areas, but those ancient codes, genetic and mimetic, began to manifest wilder than ever, sixty years after toddling off alone to the fields, at the age of four. Sometimes I felt twenty-one again; my legs and feet guided me swiftly through dark and cactus-laden nights. My reborn litheness reminded me of my logging experience (chopping down trees of three inches or less diameter) helping clear the Hungry Horse Dam reservoir's steep mountainsides in the Montana mountains in the early summer of 1950. We ate well-done steaks for breakfast, lunch, and dinner, with pancakes and eggs in the early morning, but no one gained a pound.

John doing some target practice while exploring the Front Range in the Colorado Rockies.

All my life, since I first left home at fourteen to work in the war effort in Los Angeles when all the able-bodied males had been drafted, my motto had been "create and run." Continuous creation was my dream. My attitude, until well after the Vajra Hotel had been finished, had been as if my life was symbolized by Brahma and Shiva alternating – creation and destruction. I never really went for the Vishnu super program, the maintenance force, until later, when Krishna (an avatar of Vishnu) arrived with a message traced out by his dancing feet, in a ceremony danced in the Vajra Hotel pagoda just before Biosphere 2 started. I had to leave Biosphere 2 with enough data for me and others to take the next step. I had to maintain the quality of that vital data until my part in that epic ended.

I grew up listening to the legend of my great-grandfather, Buck Wall. Buck had refused to go along with Texan slaveholders in seceding from the Union in 1861, siding with Sam Houston, our family's great hero, "The Raven," to give him his Indian name, in opposition to this disastrous move. My grandmother's family was the Polks for whom Polk County in Texas was named, and Houston had been Polk's chosen successor to carry on Jefferson-Jacksonian traditions. When told by the traitor government that he had to join up to fight the Union, Buck had said, "Come and get me if you can. I do not fight for slavery and disunion."

He paid dearly. One of Texas's ten great feuds erupted between his family and returning Confederate hard-cores after the Civil War. However, he never ran, not on an issue he perceived to be a transcendent imperative. That legend had nourished me in ways I had not realized. Slavery's time had run out; now the time to ransack and enslave the world of life and cultures as commodities had run out too. Knowledge-action was required – knowledge that Biosphere 2's Mission 1 would provide to people ready for creative action. Some would try to run down, conceal, or destroy that knowledge.

During this time, Margaret directed a production in our theater space of my adaptation of *Prometheus Bound*. Margaret played Prometheus with more than her usual flair and depth. She reached *duende* with emotional truth. The costumes and lighting dazzled me. The story is that Prometheus was severely punished by Number One for having made inventions that helped humanity against Number One's orders. Audiences from Tucson and the town of Oracle packed the Bunkhouse Theater to see the show. A Stanislavskian would say that part of its power was undoubtedly due to the fantastic subtext that Margaret and some of the other actors used in which the biosphere's creative power could stand for Prometheus, and Zeus could stand for state and corporate wreckers of Earth's biosphere.

A Turkish theater director, once drolly stated that Kathelin and I had carried the Stanislavsky method to the ultimate extreme. Our team, he joked, had created "all those madly different projects around the planet" just so we could play any role on stage. Brecht (history) and

Artaud (inner intensity) and the Hindu Nataka (physical dexterity) made up the other three of Kathelin's and my inspirations in theater.

It required daily practice of techniques learned from Brecht's distanciation and Artaud's appreciation of the theater and practice of movement, as well as Stanislavsky's emotional memory and the quirkiness and tempo of street theater, to keep my humor and vitality going at Biosphere 2. I reviewed with fresh interest all my demonstrations in acting classes. Working with friends who had also "been through the mill" took a load off my shoulders. I took the energy given by the impact of my time at Biosphere 2 to dig deeper into history and theater. What *should* I write? What *should* I do next? I didn't want to live the rest of my life as an anti-climax.

Besides the Prometheus production, on Tuesday nights we threw parties for area intelligentsia. The theme of these parties dealt with crucial periods of world history. Herodotus and the Persian Wars, Thucydides and the Peloponnesian Wars, Gibbon and *The Decline and Fall of the Roman Empire*, Parker on the fall of the Aztecs and Incas, Ibn Khaldun on the rise and fall of Arab dynasties, and on up to Churchill and closing the ring in World War II. The problems of the day were put into perspective by considering those of Themistocles, Pericles, Septimus Severus, Cuauhtémoc and Churchill.

Sometimes I went to movies, including some favorites like "Bladerunner" or "Flashdance." For sheer relief, "The Teenage Mutant Ninja Turtles" cracked me and Deborah up. Allen Ginsberg came out for a day after one of his rousing readings in Tucson. He had attended one of our plays in Germany and had bounded into our dressing room after the performance. "Where's the orgy?" he grinned when we offered him coffee backstage. Cecil Taylor delighted me with his white light piano playing. Kathelin had learned it from him and I from her, and now I got to meet the master. Ravi Shankar reminded me of cosmic perspectives and glorious times on the Subcontinent. I had met Arthur Clarke in Sri Lanka while doing coral reef research; he talked with us now and again by videophone, ever a radiant inspiration to never give up on the future accomplishments of human destiny. Isaac Asimov wrote about his intense disappointment that his aversion to flying on an airplane prevented him from seeing Biosphere 2.

William Gibson, whose *Neuromancer* joined those books that changed my life for the deeper, visited; we had a thoughtful encounter. His meticulously realized vision saw great megalopoli dominating the biosphere, whose totality revolved around black holes of billionaires with riches so dense that their concentrations formed event-horizons from which no information could escape, just as black holes formed event-horizons from which no light could emerge. Gibson enlightened my reason and energized my inner shaman. The New World Order of Neuromancer-like black holes of finance might "disappear" the whole event. Biosphere 2 information might never reach the public or world science, any more than light could escape a black hole.

Gibson's work made clear to me the mechanics of how so little information escaped the oligarchy's control of cultures, technics, and the life forms of the biosphere and how their remorseless monetization of everything degraded our potential.

Marlon Brando spent a full day with us at Biosphere 2. Taking him "everywhere," I had a chance to study his intuitive post-shamanic skill in picking up on gestures, even flickers of a gesture that revealed the essence of a character trait. A couple of times he asked if someone in one of our laboratories would repeat some special movement. His observational power of e-motion, those flashes of muscular twitch that reveal someone's inner life, equaled (if it didn't actually surpass) the abilities of Lorenz and Schultes to observe the slightest movements revealing behavior patterns set in motion by instinct or ritual. An idiopathic naturalist of emotions and moods, Marlon sized up

*Many great scientists supported the project with their experience. Here we prepare to
welcome the crew back to Biosphere 1 on September 26, 1993.* ABOVE FROM LEFT:
*Oleg Gazenko, Leonid Zhurnya, Harold Morowitz, Jane Goodall, Robert Hahn
(our communications officer), Sylvia Earle and John Allen.*

each individual that interested him as Lorenz contemplated a
species. We became friends in long telephone conversations on ecology, especially relating to whales and coral reefs. Some years later,
Marlon helped the *Heraclitus* in its planetary coral reef studies. He
died just as I finished the first draft of this book; I felt a great light
left America.

We scheduled monthly visits and lectures, open to the whole staff,
by outstanding scientists, like Howard and Gene Odum, Richard
Evans Schultes, Harold Morowitz, and Oleg Gazenko with whom
the reader is now acquainted, and a dozen others. I sat at the feet of
these masters in rapt attention. Morowitz, who had a gift for synergizing wit and philosophy with detailed science, had shown the exact
chemical steps by which a biosphere increased its free energy, as long
as it was open to energy flux. Investigating Biosphere 2 with these

masters made me realize again, as Newton had said, that I was a boy playing with pebbles on the shore of a vast ocean. Sir Isaac added that, indeed, for him to see that much, he had to stand on the shoulders of giants. I preferred sitting at their feet, where I could look into their sensitive and passionate faces, these giants of intelligence who had made biospherics and Biosphere 2 possible.

Later, when my direct relationship with Biosphere 2 ended, I began to see the next decisive step in developing the science of biospherics. While out on one of my barefooted desert rounds, I saw in a flash the next step in working out the biosphere's full range of interactions with the ethnosphere and technosphere: creating a cybersphere. This would give a complete feedback system to the people who operate the commodity-driven technosphere, Global-Tech.

This cybersphere would synergize various kinds of data, using conversations, expeditions, the internet, space monitoring, conferences, and reports from observers on occurrences indicating sharp changes in their areas. Without this synergy, no real connection could emerge between the world of disassociated information and the needs of the peoples of the ethnosphere and of the biomes of the biosphere. I staggered around my mulberry tree-lined lane like a newborn calf. My new set of futures, my Maroufian Caravan, had burst into glorious sight over the distant sand dunes. I could hear those camels bellowing now. I could hardly comprehend how elementary my Biosphere 2 project now appeared to me. Our first Ethnosphere conference took place in October 2002, the Cybersphere conference followed in November 2003.

This *vision* of a cybersphere brought me a brighter, fresher light. My life starting once again! I understood how to proceed, equipped with the great lessons and inspiration of Biosphere 2. It had sailed far and adventurously enough that its legend would survive any shipwreck that might occur. From this moment on, I would base my actions on the collective wisdom of the ethnosphere, including its new world culture of Global Tech. For the first time, I knew how to regenerate my life after Biosphere 2. It had been ten years of unremitting toil and I had been in denial of the toll from the stress of the past years, but this pay-off-beyond-compare started regenerating my very cells.

This cybersphere notion also showed me how the *Heraclitus* could apply the findings from the Biosphere 2 coral reef and thus carry out my original idea behind starting the Planetary Coral Reef Foundation in 1991, just before Mission One. Even then, I had begun to prepare a lifeboat. Shackleton had sailed a lifeboat from Antarctica two thousand miles through storm and cold to land smack dab on his tiny target island, to save his expedition. My problems were minor compared to Shackleton's. My Special Forces friends always had a deep hole dug in the middle of their compound. I had found out about it one day when a "thump" occurred. Sergeant Jackson yelled, "Incoming mail!" and dragged me down a narrow entrance into a place covered with lots of comforting earth before the explosion.

The Viet Cong had caves, but they weren't the only ones. My cave was my subconscious rooted in the biosphere's evolutionary powers; my lifeboat was my ability to think anywhere – and now I had found out how to contact those creative forces directly: align my being with the forces of new life in *every* culture. How had I been so dense not to have seen this before? The ethnosphere was my Finagle's Factor.

CHAPTER EIGHTEEN

Mission One Victory

*There are two kinds of scientists; those who see Biosphere 2,
and those who don't.*
JOHN CORLISS, DIRECTOR OF RESEARCH, BIOSPHERE 2
Time, 1993

EIGHT HEALTHY, GLOWING BIOSPHERIANS EMERGED on September 26, 1993, after two years sealed inside Biosphere 2. "They are well on their way to establishing a world-class center for excellence for the study of ecology," wrote geneticist Steve O'Brien in the Phoenix Gazette on September 19, 1993. On September 26, 1993, Dave Chandler at the *Boston Globe* summed it up: "Their experiment has proved the feasibility of a totally self-sufficient enclosed system." "The public loved it," wrote Gerry Soffen in the *New York Times*, in February, 1994. The great systems ecologist, Eugene Odum, wrote in *Ecology* in 1997 that "Viewed as an exercise in human ecology and environmental engineering, the experiment was a great success." Howard T. Odum wrote in the journal, *Ecological Engineering*, "The management process during 1992 – 1993 using data to develop theory, test it with simulation, and apply corrective actions was in the best scientific tradition."

Harold Morowitz said at our 1992 Closed Ecological Systems and Biospherics conference that "Biosphere 2 provides, for the first time, the possibility of conducting controlled, large-scale ecosystem ecology experiments. Modern physics emerged when Galileo conducted experiments that yielded numerical data. Biosphere 2 enables the same type of transformation to occur in ecology as occurred in physics."

This fantastically successful finish of the unprecedented two-year Mission One succeeded in proving that complex adaptable life systems

IMAGE ON LEFT: *Reentry of crew to Earth's biosphere on September, 26, 1993 in front of thousands of onlookers and the world media completed the longest human enclosure experiment, by far, ever conducted.*

can sustain a long-term, healthy, creative human culture, in a harmony with a high-tech system which in turn can enhance the life systems. Thriving biomes can be designed, built, and operated, and be restored to original habitats even when they've been heavily damaged. The public can be easily educated in complexities if people are allowed to see real-time large-scale science in action.

New inventions were spun off the invention of Biosphere 2. Eleven patents, the complete recycle of water, food, and waste, and a less than three hundred ppm a day atmospheric leak had silenced all the doomsayers. Thousands of well-wishers traveled from far and wide to celebrate the biospherians' re-entry, the first voyagers into another biosphere. As soon as I could (after the biospherians left to go through their examinations and to their meetings with friends and families), I dashed inside to my favorite spot, the one specially included for biospherian system contemplation. I had lots of experience, needing con-template-ing, to digest.

Standing alone on the high balcony of the biospherians' micro-city habitat, I lingered over this scene, pure space-time poetry to me. I had conceived and Margaret had designed it to give a taste of looking over the fields of Tuscany from a great villa at the edge of Renaissance Florence, to me the epitome of civilized living. Taking deep breaths of the stimulating non-polluted air whose composition I knew to the last detail and how it cycled during days and nights and seasons and why, I then drank my fill of water whose exact composition I knew and the exact paths it followed in recycling to arrive in my glass, more delicious and stimulating than the finest wine. Agricultural plenty spread out below me, in checkered fertile fields of diverse and productive cultivars, from rice paddy to bananas to peanuts to sweet potatoes and wheat.

I knew the soil composition, climatic cycles, insects and cultivars, what carbohydrates, proteins, fats, and trace minerals each cultivar contributed to human physiology, the amount of labor and care needed to produce and maintain each component of this system. Soils, the main support of human life, are the least understood of elements in the biosphere. Their dark world, swarming with immense numbers of microorganisms, worms and mycorrhizae, challenge the most ingenious observers and data collectors. Their interactions with the atmosphere, exchanging gases, create an incredibly intricate dance yet to be deciphered. The agricultural biome stands at the apex of the field of necessary biospheric experiments.

Behind me, on the ground floor to the right, in the animal area, lived the small, hardy, productive goats from the Jos Plateau in Nigeria and the Rhode Island Red chickens similar to my grandmother's reliable providers of meat and eggs. Like in many European villages, these animals lived under the human habitat. The humans, those political animals, as Aristotle described us, could easily descend in the morning, greet their evolutionary companions, feed them, milk the goats, collect the eggs, then walk straight out to fertile fields whose green chlorophyllic wizardry fed all of them.

Working the agricultural system, plus food processing, cooking, and waste recycling, took forty percent of biospherians' work time. That satisfying, necessary process would attune anyone to reality. This union of intellectual, manual, and sensory work would end anyone's specialist-caused separation from the fountains of life. It would unite the human organism in emotional-aesthetic rapport with the harmonious beauty, picturesque romance and sublime setting of their spacious architecture.

Behind and above me, along a corridor, Margaret's elegant tower soared to a lofty dome. Here one could browse in a library filled with the writings of thinkers from around the ethnosphere or peer through a telescope far out into the cosmos, or take in a sweeping view in which a biospherian could contemplate two biospheres and the stars. Here one could celebrate the cosmos through which they revolve, rotate, wobble, and spiral in their glorious home, Planet Earth, its cycles materially closed by its gravity. Through the glass floor of the library, one could gaze straight down sixty feet to where biospherians brought crops in from the fields and whose loving transformation into cuisine linked contemplation and sweat.

In that book-lined dome I gazed out through the Platonic-Fullerian-Augustinian geometry of glass walls and ceilings to rugged mountains and deserts and up into blue light-laden sky. Rolling my eyes to the left, I caught glimpses of the great continental biomes: tropical desert, savannah, and rainforest. The graceful double-inflected domes of the lungs flanked the biosphere, protecting its atmospheric pressure while automatically maintaining equilibrium with local weather vagaries. Refreshing waves lapped over the coral reef on the other side of the savannah, below the plunging cliff; the dark, life-filled mangrove marsh lay beyond the coral reef. I looked forward to strolling down the sandy beach stroked by its gently wafting breeze, and taking a meandering swim through the symbiotic corals.

I felt that this moment's taste of the harmony of technics, life, and culture intimated how life should and could taste for all humans – at home in their cosmos, their biosphere, at peace with their destiny.

I ran up steps two at a time to the all-electronic offices of the great command center, softened by one hundred percent natural wool rugs and wall panels donated by the Wool Bureau, with its state-of-the-art computers donated by Toshiba. Each biospherian had access to video, computer, and telephone communications, which allowed them to stay in contact with our scientists, artists and their friends.

I looked in on Roy Walford's thoroughly equipped and simple to operate medical laboratory and thought how every neighborhood should have one of these. What a practical way to re-introduce the general practitioner of old, the doctor who knew everybody but who also possessed all the modern techniques and apparatus right there in his/her own shop.

I looked in on Mark Van Thillo's machine shop in which everything mechanical in the technosphere of Biosphere 2 could be repaired. It was a recurrence of the ingenious Yankee garages of Henry Ford or the Wright brothers.

I grooved on Taber MacCallum's spotless analytical laboratory. Every city neighborhood and village ought to have one of these, too, and end its dependence on some remote bureaucracy. Then, one could quickly measure any change in one's water, air, food, soil or other chemical vector.

I sat down at the polished granite table in the dining hall, then gazed straight into the rainforest with its waterfall, which I had modeled on those in Guyana (where Sir Arthur Conan Doyle had located his Lost World), a place highly favorable for evolving new life forms.

I visited the theater space where the biospherians had danced, made videos, played music, and celebrated. Then and there, I made my own Dionysian dance to Biosphere 2.

I could hold back no longer. Moving very quickly now, I descended into the technosphere, placed underground in order not to waste valuable sunlight that could be used by the biosphere. I followed the long pipes carrying water, emulating the underground streams and caverns of the Earth. I reached the marvelous cave-like lungs that Bill Dempster had designed, that expanded and contracted beneath their serene domes, keeping the inner and outer atmospheres in perfect balance, preventing any explosion or implosion resulting from pressure. There, standing under seventeen tons of steel and rubber riding up and down solely on air, was one of the perfections of engineering. I stood silent at the edge of the underground lake that pro-

Biospherians through the glass window on the morning of re-entry to Earth's biosphere, September 26, 1993.

vided a reservoir for Biosphere 2's hydrosphere. This recycled, pure water was the heart of "biospheric blood circulation," as Vernadsky saw the hydrosphere.

Satisfied that the vast technical system was still in working order after the two-year experiment, I dashed back to the rainforest. I climbed to the top of the rainforest waterfall to breathe the pristine air of the cloud forest, then slowly climbed down the other cliff to the várzea (riverine) bioregion. I strolled through the ginger belt and finally burst out into the savannah with its luxuriant grasses and its riverine forest. Then, I paced carefully onward through the Mexican thorn forest, a bioregion that always transports me into shamanic states.

I glided into the desert. It had been included despite the fact that its low biomass and slow turnover rate meant that it did not contribute much to Biosphere 2's flow of materials and energy. Nonetheless, deserts play such an important role in the evolution of Biosphere 1 that I refused to leave them out. For one thing, they have contributed so much to human thought and art, and for another, it's essential to measure their interactive contribution to the other biomes to obtain a real model for biospheric understanding.

With increasing excitement, I made my way through the flourishing mangrove marsh biome, remembering adventurous times in Belize and Fiji, lost in the tangles of those magnificent root systems. They can prosper in such a wide range of salinity, and they host a multitude of life forms, especially all-important invertebrates.

At last, I treated myself. I swam around and about the coral reef ocean. I had helped carefully hand shape the final contour of the beach. I had experienced so many adventures with the *Heraclitus* searching out the right place to collect corals to model this biome! This reef had required perhaps the most careful measurements, observations, and management of all the wilderness biomes, and it had survived and increased although several experts had predicted its rapid demise.

I had modeled this wilderness ecoscape on vivid memories of my great African walk in the first months of 1964. Biosphere 2 recapitulated ecosystems from the mountains of the moon, down the escarpment to the savannah, to the desert, to the marshes and coral reef around them, the then still dhow-rich old Swahili port of Mombasa where *Heraclitus* had anchored. One of my heroes, the great explorer Richard Burton, had many adventures in those ecoscapes; on its edges, the Leakeys had discovered the origins of humanity to be not from the apes, but from offshoots of the more gracile *Australopithecines*.

In September 1993, biospherics began to hum even more, after the dramatic "re-entry" of the biospherians. The research, development,

John Corliss, who led the re search project that discovered the first thermal vent ecology at ten thousand feet under the sea, joined the SBV team in 1993 as director of research.

and engineering team was much stronger than when I had started it back in the pioneer days of 1984 and 1985. In addition, naturalist John Corliss, one of the discoverers and observers of hot vents on the ocean bottom and a stimulating thinker of the first quality, had now begun as director of research; he was rapidly picking up on the complexities. Don Spoon, the microbiologist from the Smithsonian, had also joined us. Besides sharing his invaluable observations of the tiny ones and their manifold ways, Don taught me about irises and their world of ever-increasing beautiful varieties. Margaret and I added some of his gorgeous plants into our ecoscape.

Consulting rather than directing became my econiche. I stayed in daily communication with Bill Dempster. My research, development, and engineering workload was reduced to walking around the action area with my associates each morning to check hot spots and review the intellectual terrain. This gave me time to think strategically about biospheric enterprise and what my new work in cyberspheres and the ethnosphere might be after I left Biosphere 2.

After re-entry, biospherian Mark Nelson plunged back into working with me on theoretical aspects of biospherics. He could now participate fully again with Ed, Margaret, Kathelin, Marie, and me on the board. Mark had honed his formidable political skills by running the Institute of Ecotechnics for years. He began to work on obtaining his doctorate in wastewater treatment ecologies.

A completely tested managerial team now covered biospherics in business, building, maintenance and operation of Biosphere 2 and managed seven new biospherians during the all-important Transition Mission. Marie ran the numbers with a new zest. Ed Bass had delegated a lot of his work as director of finance to others, but he continued ably to review, analyze, adjust, and approve the monthly budgets. The SBV board began discussions as to when and how Margaret, Marie and I would leave. I needed a year at least, probably two or three, to re-think and re-experience Biosphere 2 and our decade of on-the-edge creativity. Would I return to big projects or should I become primarily a writer and contribute to the on-going ecosystem projects like Birdwood Downs in Australian or the rainforest project in Puerto Rico? How could my new insights on the ethnosphere and cybersphere be translated into practical action? I needed meditative time; I needed some aloneness in an orchard, a jungle, a savannah, or out on a remote reef with the *Heraclitus*. I needed time in roaming libraries of the October Gallery and Synergia Ranch.

I increased my contemplation time on desert hikes and on occasional horse rides with Clare St. Arnaud, a student of Crazy Horse, active in the struggle for Indian rights. He became a triathlon Ironman to help inspire Indian youths not to drink. We rode the rugged back-country, back and forth from Sunspace Ranch to the lively artists' community in Oracle, led by the insightful Anne Wooden and genial Andy Rush (who did the wilderness drawing for the cover of my book, *Wild*). One time I rode out of the bush with Clare, Deborah and Margaret and dismounted in the Oracle artistic community to read my poems to a small local crowd. I had published *Wild* in 1992. It included some poems about the Biosphere 2 experience, aphorisms and short stories from my early travels as what Jack London called a "bohemian of the working class," otherwise known as a migrant worker. One aphorism was "Life is the creative imagination of the dead." One aphorism I didn't publish, "Death is where creative imagination is all that's left."

I began to see how to complete the final volume of my Joe Madison trilogy, *Liberated Space*, which dealt directly with the Haight-Ashbury period of the 1960s and explored the cosmic network of meanings that had beckoned many of us. I gave well-attended readings at the Caravan of Dreams in Fort Worth, of my own poems and those of others. At one reading, I acted as Lorca reading his own work. "Green, green, I want you green." Verde, verde, te quiero verde.

The BRDC greenhouses where early experiments were conducted to select different plants for the agricultural biome became part of the Visitor's Center so people could see and experience something of what it was like for the crew to live inside Biosphere 2.

BRDC greenhouses housed a small animal bay with chickens and goats, an example of the fish-rice-azolla system, and a large area representing the rainforest biome.

To my surprise, I loved helping Margaret develop the beautiful Visitor's Center, where people from all walks of life could taste real-time science. They could look through the same windows that our consultants and I had used, and make their own notes on the progress of the biomass in the diverse ecosystems. They also had full access to all our research biomes. A trip to Biosphere 2 was a prize sought by the brightest students in California, Arizona, Colorado, and New Mexico. Busloads of students arrived in the spring of 1993 to get full briefings on ecological systems. Fourth- and fifth-grade students were inspired to urge their teachers to let them build model biospheres in their classrooms. Millions of people tuned in to our educational programs.

When I had traveled through the research biomes with that wise ecologist Larry Slobodkin, he said, "I wish I had these for my experiments." A number of scientists told me the same thing. We had made many friends by giving scientists and artists full access to the project. Big names could anonymously tour through BRDC and Biosphere 2 and could make up their own minds about it. It was in that way that we met open-hearted and razor sharp Frank Salisbury, an outstanding plant physiologist and experimenter with wheat in NASA's closed systems, and a friend of Galina Nechitailo. They had run experiments with wheat aboard the Mir Space Station. In Russia, on the way to the second International Conference on Biospherics (held in Krasnoyarsk, Siberia in 1989, and co-sponsored by Institute Biophysics and the Institute of Ecotechnics), our bus broke down. Frank immediately had us walking with him through the fields, expounding on the plants and ecology. Almost with regret, we saw the bus driver waving to us to return. Frank became a close friend and scientific colleague and an invaluable source of critiques. He did the remarkable editing of Vernadsky's *Geochemistry and Biosphere*, published by Synergetic Press, in 2007 – a labor of love and intelligence that allows the English-speaking world to experience for the first time the very best of Vernadsky's beautiful, complex thought.

Frank Salisbury speaking at the Fourth International Conference on Closed Ecological Systems and Biosphere Science, held at the Linnean Society of London in 1996.

We reviewed the correspondence of patterns in year two with year one. The chief reason I had picked two years to be the length of the experimental Mission One was to better make improvements for experiments in future missions, which would last about one year. Checking annual patterns (because they are objectively based on solar years), is the best way to find a control equivalent in time to a control in space. That's how all great observational sciences advance – astronomy, climatology, planetology, biospherics, and my new passion, ethnospherics. For the moment, all seemed well. Still, I resolved to extract all the life, friendship, and knowledge possible; I sensed there were only a few more months.

Biosphere 2 would become a mystery drama where Dionysius and Demeter (Theater and Life) would disappear into Legend. Myself and fellow actors, scientists, and explorers would strike the set at the climax.

Cherry Blossom Time

CHAPTER NINETEEN

Cherry blossoms fall.
Dreamed actions and acted dreams.
Emperor's gardens.
JOHNNY DOLPHIN
IMPERIAL GARDENS (TOKYO, JAPAN)
HAIKU, APRIL, 1994

WHILE CREATING BIOSPHERE 2 AND DURING MISSION ONE, I was invited to visit Japan three times. On two of those trips, Margaret Augustine, Mark Nelson and I gave talks and visited technical facilities. We had welcomed a number of Japanese scientists, and Mr. Toyota himself, to Biosphere 2. Mr. Toyota inquired as to why we didn't have a Toyota inside. Japanese space life support system scientists like Dr. Keiji Nitta, who had been placed in charge of creating their closed life system (Biosphere J, now called the Closed Ecological Experimental Facility), fully grasped the implications of Gitelson's and our work in terms of space settlement. One Japanese construction company had offered Margaret and me $150,000 to start up a small joint venture. The Japanese pursue moon studies with the same seriousness that a section of NASA pursues Martian studies. To live on either Mars or the Moon requires a mastery of biospherics.

After the decisive accomplishment of Mission One, interest in our knowledge of biospherics and in its practical applications reached an all time high, nowhere more than on that island empire of Zen and science. I made appointments with heads of major corporations who had expressed interest in applications in previous meetings and with Keiji Nitta.

Deborah Snyder and I were met in Tokyo by two extraordinary men, Gessie Houghton and Divyam. Gessie, a versatile linguist from

IMAGE ON LEFT: *Tokyo, April 1994, promenade outside the Imperial Palace.*

Cambridge, had spent fourteen years teaching at the University of Kyoto, and had become a top martial arts specialist and a director of his master's elite school. Divyam was a French ex-stockbroker who, after leaving Paris to study metaphysics in India, married a beautiful world-traveled Japanese woman, becoming an expert on Japanese art and a philosopher. Gessie and Divyam were to join us in meeting John Lilly. They, like me, admired John Lilly's work with dolphins. John's isolation tanks allowed sights, sounds, smells, light, gravity, and temperature variation to be eliminated. John thought that what one then perceived would be coming from inside and, properly used, this technique could unravel secrets of human metaprograms.

Our new friends took us first to Divyam's lovely residence on an old cobbled street that the Emperor used to stroll along between the Shinto-flavored graveyards and the Zen temple that stretched up from the opposite end. Then Gessie drove Deborah and me to the house where we would greet John Lilly.

On the morning of April 2nd (April 1st in the U.S.), the telephone rang. Marie Harding said that the expected negotiations between Decisions Team and Decisions Investment corporations had been initiated by DI for the dissolution of the Space Biospheres joint venture. Marie urged Deborah and me to continue with our Japanese visit. She was ready to head up these negotiations for the transfer of responsibility of Biosphere 2. Marie would contact me twice daily by telephone. She, Mark Nelson, and Robert Hahn would hold daily meetings. She would telephone Margaret, would deal with our lawyer and prepare the negotiations with Mark. I felt Marie could handle the situation since we had been discussing together options for the past year. She and I had handled many a difficult situation since first working together at Jim Turpin's remote medical camp on the Ho Chi Minh trail, in 1964. She knew the business.

I wandered through the cherry blossoms near the Ryoanji Temple that first day. Phone calls with Marie and talks with John Lilly, Philip Bailey, Deborah, Divyam, and Gessie pulled me back from vertiginous samadhis reviewing past changes in my life's existential direction.

One memory took me back to the U.S. Army Corps of Engineers, in February 1953, with my fourth echelon unit building an Air Force base near Orlando, Florida. I had been transferred there after I had finished advanced basic training at Fort Belvoir. I had just received a message stating that I would receive a general discharge in three days for "suspicion of subversive activities."

In that case, suspicions arose from the fact that in 1950 I had been elected secretary of the Anti-Discrimination Committee of District 10 of the United Packinghouse Workers of America, in South Chicago. Two of my great mentors there had been Paul Robeson and W.E.B. Du Bois, the brilliant forerunners of the moral grandeur reached by Martin Luther King. For his unsparing tearing off the veils of the crimes of segregation, the State Department no longer allowed Robeson to travel abroad. The Communists, for their own purposes, also concentrated on this weak link in American society. Both Robeson and Du Bois found the South Side of Chicago the only free area of movement left to them in America.

Robeson had given me several priceless lessons about real voice, which reinforced my love of theater. I remembered him singing "Old Man River," raising two thousand workers to their feet at the Packing House Union Hall in 1951 with a single upthrusting gesture of his powerful fist. Robeson changed one line a bit: "Show a Little Fight, and You La-a-a-and in Jail! But Old Man River, that Old Man River, he just keeps a-rollin,' he just keeps a-rollin,' he just keeps a-rollin' along." Paul kept going, no matter what changes occurred.

That unexpected general discharge "on suspicion of subversive activities" left me somewhat fearful. Maybe some zealous shavetail intended me to get roughed up a bit by patriotic soldiers. The war in Korea still raged. I walked out of the captain's office with my notification and, naturally, everyone had already been tipped off. A few big guys were waiting for me. I sighed and readied myself. One guy came up and said, "If you're a Communist, Johnny, I want to join up." "Me, too," said the other guys.

My faith in Americans being able to see through malicious bullshit was reaffirmed. My sigh of relief, though, was audible to all of them. Later, under the enlightened rule of Jack Kennedy, I and over four hundred others who had been given the same high-handed treatment were allowed to meet one by one with a review committee of "four birds" (lieutenant colonels) and one full colonel who voted that my discharge was wrong. They made out an Honorable Discharge with my sharpshooter and other badges intact, and said the other discharge no longer existed.

That memory and others flashed back to the various times my life had been drastically changed. I realized that all these moments had brought greatly heightened intensity of thought, vivid sensations, and the thrill of being alive.

In Tokyo, cherry blossoms fell around us as the future of Biosphere 2 was decided.

I could only imagine what Deborah, my colleague and companion, was feeling on our journey through Japan, since she had given her straightforward and magnificent all to Biosphere 2. She was loved by everyone, and created a series of beautiful books about biospherics. I had planned all along to leave my position on my sixty-fifth birthday, May 6, but I was finishing my task and she had just started her big project. I had already roughed out my next steps, but now she wouldn't know her situation until the negotiations finished. She behaved like a seasoned veteran and never lost her cool. We took a few days to enjoy Japan in the best of company.

For decades, I had wanted to explore Japan during cherry blossom time and I decided to use my new leisure time to take them in, in all their glory around Ryoanji in Kyoto, at the Philosopher's Walk, and also in the Imperial Gardens in Tokyo. I had been told that in each cherry blossom falling red upon the ground, the Japanese see a young Samurai dying in his first battle. Tears flowed out of my eyes when I arrived on April 5th at the blossom-spattered Imperial Gardens. With every fallen blossom, I saw species dying and biomes being degraded. That storm of blossoms sank my knees to the ground and I was racked by sobs for Biosphere 2.

The Sufis say that every obliteration reveals a new confirmation. I realized that the miraculous was happening to me. My search had succeeded. At sixty-five, on the verge of retiring, this rare, unforeseeable and astonishing birth of a new life arose; once again, the great teacher Kali had transformed my social identity. Health flooded through my cells, adventure through my nerves, ideas through my brain. All the ninety-eight emotions I had identified for our theater tumultuously renovated my hormones. The nine moods picked me up and threw me into their nine different universes, one after the other. I saw Kali herself, in direct vision, her mission as my teacher finished. I first saw her in Kathmandu. Kali, Mahashakti, The Dark Lady, Ishtar, Mary Magdalene, the Great Mother had moved on after giving me really good lessons which vivified the unforgetting cellular level.

Statue to the goddess Kali, Kathmandu, Nepal.

After returning from Japan, I stayed in close telephone contact with Marie, but decided to lay low and have some fun on the side – to start decompressing from the decade deep dive into Biosphere 2. I went to Los Angeles where Chuck Daugherty, my old Colorado Mines football and metallurgical buddy, bought me a small fax machine so I could get back into written communication with Marie. He lived in the San Fernando Valley where he was doing engineering work on the Space Station.

Chuck took me through the model of the Station and introduced me to the engineering staff, all of whom were highly interested in Biosphere 2. Oscar (called Oz because of his wizardry with the psyche) Janiger invited me to stay for a week in his idyllic Santa Monica home. His library could entertain and instruct me for years with its fifteen thousand masterpieces, most of them on little known but highly rewarding aspects of the human phenomenon.

On the great sandy beach near Oz's hidden garden, I vividly remembered turning fifteen and arriving in California, when I first set out to see the big world on my own. I discovered Santa Monica beach in 1944, when I thought myself very lucky to land a job working forty-eight hours a week, for sixty cents an hour, at Helms Bakery. But I could take a girl out for the evening for five dollars and could study dancing at Fred Astaire's studio for a couple of bucks. I had been making four dollars for fourteen hours of work every Saturday on Double Tough Street, back in Oklahoma. Double Tough Street's movie theater featured Grade B Westerns and was famed for some holes in its silver screen where an occasional member of the audience shot at the villain when the hero seemed too feckless to help himself!

I spent Sunday mornings on the beach, breasting the waves while watching for the undertow, then lounging about while soaking in the striking beauties and the rippling muscles. Blonde seventeen-year-old Ola, one of its beachcombers, charmed me out of my mind. Ola dreamed of finding a man who, when wildly jitterbugging with her, would know exactly how to flex his hand on that magic place she believed existed in the small of her back. This would, then and there, transform her into Ginger Rogers! I assiduously experimented while she helpfully snaked her back as we danced on the pier. We found that perfect spot and touch, though we could never keep it up at full throttle longer than for half a Harry James Orchestra number. But what great fun! How we whirled and swung and dazzled ourselves.

After Sunday morning swimming and sunning on the beach, I would walk up the steep path to the palm-shaded palisades and meet thoughtful men, exiles from fascist Munich, Prague, Rome, Berlin and Vienna (names that transported me with romance and the hope of knowledge). They played chess in a small, airy, green-roofed building. The ruthless events that had flung them across an ocean had been the takeovers of their nations by dictators.

I had read Tarrasch the year before. "Chess, like music, like love, has the power to make men happy." Because of my quoting Tarrasch, these exiles let me be only the second American ever allowed to play them. The other American was a khaki-shirted, one-armed, tobacco-chewing checker champion from Arkansas who was enough of a philosopher to intrigue them. Tarrasch's teaching revolved around tempo. A gain of tempo on your opponent could be worth as much as a knight. This "extra knight" would enable you to checkmate the enemy. His teaching was that, on occasion, you can sacrifice tempo (and a pawn) to secure your position. Position can be worth as much as a rook. One gemütlicher German gave me a chance to gain tempo and a pawn. I could not restrain a grin of triumph. "Bah," he smiled, "what's a pawn amongst friends?"

David Jay Brown, who had written an enlightening book, *Mavericks of the Mind*, stopped by Oz's to include me in his next book of interviews, *Voices From The Edge*. David's interview gave me a chance to record some of my views on biospherics, while still in the ultra-high energy and conscious state of "raptures of the deep."

During my time at Oz Janiger's, I found myself rolling, moaning, and pounding the floor in a regression that proceeded back to a mythical moment of conception and re-imprint of my life. Oz joked in his avuncular way, "Johnny, I thought you had gotten past free-floating anxieties."

Wise "seen it all" Oz and relaxed "hide it all" Santa Monica gave me a bibliophilic and nostalgic week off from the front, save for the twice-a-day conversations with Marie. Acting the professional psychologist, Oz prescribed a hefty dose of KGB truth serum and told me to interrogate myself and expose my inner traitors. "Those moles dig really deep," he said. I didn't find any inner traitors, but I did find a point of delight in contemplating the unity of dualities, one now yin and then yang, and the other the reverse.

Fully refreshed, I rolled back to Tucson via Greyhound, to work side by side with my pals again. My dream of pals resumed that April, when the seven of us original eight, plus the five who ran our projects abroad, all negotiated together with the same spirit with which we had built and sailed the *Heraclitus*. Kali guru had probed me past my depths, hurled me out to the empty blackness beyond, but somewhere out there beyond the metaprograms, beyond even the magic mirror, I survived that skull kiss and returned newborn and as fresh as a daisy.

I settled down in a small apartment to take part in the last week of negotiations. A twenty-minute walk and I could hike into the desert and fuse with a power snake cactus to restore my inner life. A thirty-minute stroll in the other direction, back and forth to the supermarket across the highway, and I could get a *New York Times*. The *Times* played fair with most things, if you knew their rules. They followed a complex party line and subtly doctored the news to fit it, but they never changed the party line in their politics and always used the same snake oil. My joke was that they always threw an undoctored bone on the bottom of page fifteen, in section A1, to keep a small econiche of people like me coming back. I called it my truth fix.

Marie had risen to the occasion, shaken off her well-founded fears and had become a skillful leader. She stayed in command of her new role in day-by-day negotiations with lawyers. The great advantage of a synergetic style is that whenever one leader or two or even three go down, a new one arises. Marie rapidly became not only a leader but the sparkplug of our new role, heading up our legal team. We argued fiercely, delighting in the process of deciding which path to take at each crossroad.

Deborah kept information flowing to all of us and acted as an ever-ready chief-of-staff to Marie. Mark brought his street smarts and controlled wrath to our deliberations. Rio kept his explorer's humor, which gave me a real kick. Teresa, Rio's companion, conjured up a marvelous banquet at her house. After toasts and Rabelaisian laughter at the follies and illusions of this world, we decided to form Global Ecotechnics, as the successor to Decisions Team and as our small rebuttal to Global-Tech's biospheric degradation. Decisions Team died at that night's party to be reborn with a more profound, comprehensive strategy and sharper tactics to work on biospherics. Kali had destroyed my old world, leaving my friends and me free to create realities even nearer to our hearts' desires. I made a tumultuously acclaimed toast to the new century, which would be the century not of the Communist-Fascist-Democratic-Republican-Labor or Conservative Parties, but of the Party!

Ed and Marie patiently reviewed the hundreds of common actions point by point since our first joint venture in 1978. After several weeks of negotiations, I considered the decision to form Global Ecotechnics Corporation had improved our historic long-run position – we could make a new start at some of our old sites. Ed Bass had been an honorable and thorough negotiator, as always. He had gained the support of Michael Crow and Wallace Broecker at Columbia University to assist in the continuation of Mission Two; they would soon change use of the facility away from its original design.

By the end of May 1994, Space Biospheres Ventures took the sixth exit. After completing Mission One, followed by the six-month transition period where we made extensive measurements and technical upgrades to the facility, and after the launching of Mission Two, the joint venture was dissolved in an agreement reached between Decisions Team and Decisions Investment regarding the future of the enterprise. From June 1, 1994, Decisions Investment took complete charge of Biosphere 2.

I felt that true world history, as distinct from the history of local tribes and empires, had just begun, with space launches, the Internet and Biosphere 2 – for the first time the entire Earth system could be examined from the cosmic reality of Space Biospheres and Total Systems Information. Naturally, it takes some time to grow from conception to birth.

The ecotechnic projects that had contributed low-cost consulting for Biosphere 2 kept operating very well during the negotiations. Chili Hawes put on a new show of a young Ethiopian artist, Elizabeth Atnafu, at the October Gallery. It got great reviews in the *London Times* and the *Guardian*. Keith Runcorn kept up on things by visiting the Institute of Ecotechnics' offices in London. Christine Handte, the *Heraclitus* expedition chief, performed admirably during all the uncertainty; she became a key member of our new long-term biospheric team. She stayed on top of the coral reef work that had supported research at Biosphere 2 and prepared the *Heraclitus* to sail to Venezuelan reefs before hurricane season. Robyn Tredwell had Savannah Systems paddocks doing well as the eight-month-long dry season began there. Kathelin Hoffman finished a new season of theater at the Caravan of Dreams, with wide support from the Fort Worth community. The Puerto Rican sustainable rainforest project did need my help to repair the previous year's hurricane damage to the single lane road we had wound into its interior, so I got out my hoe and rubber boots on my visit there.

That year, Dick Schultes nominated me to be a Fellow of the Linnean Society of London, seconded by Ghillean Prance, then

Speakers at Linnean Society meeting in 1996. (Standing from left) D.J. Beerling, Hefin Jones, Craig Litton, Ray Collins, Mark Nelson, John Allen, Galina Nechitailo, Nicholai Pechurkin, Frank Salisbury, Roger Binot, Ganna Maleshka, Alexander Mashinsky, Sergei Zhukov. (Seated from left) Tsutomu Iwata, Seishiro Kibe, Evgenii Shepelev, Abigail Alling, Phil Dustan, Sally Silverstone, Andrew Britton. (Absent) Sir Ghillean Prance, Bill Dempster.

director of Kew Gardens. The Linnean is the society I think most rooted in biospherics. Its library and quarters make wandering scholars of life feel right at home. Its learned general secretary at the time, John Marsden, was a gentleman of impeccable manners and incisive intellect, at home in the universe. Keith Runcorn and Bill Challoner, the head of the British Geo-Biosphere Program, proposed that I set up a conference on Biosphere 2 at the Society. I told Ghillean that I was not a specialist botanist or taxonomist but was looking at the way all ecological systems and incoming and outgoing energy flows work together to make a biosphere. He replied that, "The Linnean dedicates itself to studying the world of life. It needs to keep up with systems as well as systematics."

In 1996, I scheduled a conference at the Linnean Society, attended by Japanese, Russian, British, and NASA scientists. The papers we delivered there, printed (after peer review) in *Life Support and Biosphere Science*, introduced the achievements of Biosphere 2 to the scientific community. In 1999, *Ecological Engineering*, the number one journal in this field of research, devoted an entire issue to Biosphere 2 because of the significance of the experiment.

The ideology of the world order of huge banks and corporations proves extraordinarily profitable to a few, though disastrous to many, when relentlessly applied and backed by state power. Each square mile of virgin tropical hardwood can earn a timber company up to one hundred million dollars in sales, and their local official smoothers -of-the-way who signed the contract giving away the ecosystems of the tribal peoples can get a hefty cut of those dollars. Skyscrapers rise in Singapore and Sao Paulo from these "pieces of the action." A tribe might say, "It's ours and has been ours for a thousand years or more." Officials, backed by the police, would smile, "Where is the contract proving you bought it?"

When I was nineteen, I remember seeing a bright fellow of a chipmunk across an arroyo in Baja California. In the bed of my pickup, I had stashed my tarp, mattress, Navajo blanket, Coleman stove, axe, shovel, tool kit, water barrel, canteen, binoculars, typewriter and reams of white paper on which I was determined to outdo Hemingway. An inspection would also reveal a gleaming, properly polished and Neatsfoot-oiled Krag 30-40 for long range, a Marlin 30-30 for medium range, a Smith and Wesson long-barreled .38, a .410 shotgun, a throwing hatchet, and a Bowie knife. After idly pointing my .22 pistol toward the chipmunk, I pulled the trigger. The chipmunk dropped. The gun emitted a faint acrid smell and felt like doom in my hand. My stomach fell away and my head dropped to the ground. I had committed an irrevocable deed. Later on, in the army, I won my sharpshooter medal and daydreamed on occasion of using my skills in combat. But for the grace of the cosmos …

Well, I led the whole Biosphere 2 project from 1984 to 1986 until we completed the initial design. After that, I concentrated on the science and engineering. Margaret Augustine led the project until completion (from 1986-94) and made Biosphere 2 for the world. Her intelligent audacity built the complete marvel. I had envisioned Biosphere 2, examined and approved every aspect of the design and execution, as Ed Bass did from a financial point of view. But a native shyness would have kept me from pushing it to the height of artistic and scientific completeness that Margaret achieved and Ed consistently backed in his areas. I wouldn't have dared to realize my own idea to that last degree without her decisiveness and his thorough check-outs.

No one can or has ever tried to attack Margaret's architecture, since the visible appearance of her buildings refutes any attempt to downgrade her achievement. That Margaret kept this going until the final success of starting up Mission Two entitles her, I believe, to the highest status of us three co-founders of Biosphere 2. It should be known as the "Margaret Augustine Biosphere 2."

Here I would like to comment on one definitive finding from Biosphere 2. Build-ups of trace gasses are the special danger in biospherics, including our own Earth's biosphere. Chemical build-ups move more slowly, with far fewer "visuals" than physical catastrophes, but the results can be equal to or even more destructive. Molecules spread through not only the air and water, but also through soils, and they concentrate where conditions are favorable. All three vectors interact by exchanging molecules, aiming toward equilibrium but never reaching it.

For example, nitrous oxide is one of the several powerful compounds in biospheres that must be closely monitored. It will increase in a biospheric system unless a portion is split into nitrogen and oxygen by incoming ultraviolet rays in the stratosphere. Ultraviolet rays are not part of the normal biospheric loop. It is incoming energy, so Earth's biosphere, contrary to the Gaia hypothesis, does not self-organize all its nitrous oxide. Just as we had to make a wave machine for Biosphere 2's ocean, as it was too small for the Moon to raise its tides, so, without sufficient elevation for nitrous oxide to reach incoming high-energy ultraviolet rays, its removal had to be chemically or physically engineered.

A small amount of nitrous oxide is increasing in Earth's atmosphere each year due to use of nitrogen fertilizer, such that the rate of splitting of the molecules in the stratosphere no longer keeps the amount (relatively) constant. Nitrous oxide contributes to the greenhouse effect as do methane and carbon dioxide. Perhaps because this knowledge might threaten the profits generated by overuse of nitrogen as fertilizer, the media concentrates only on the carbon dioxide build-up. The U.S. Congress and President refuse to act decisively on even carbon dioxide restrictions lest the oligarchs would have to clean up their act and the air threatening our health. Only carbon dioxide and carbon fluoro-chlorides are seriously monitored on a big scale, and even then in widely separated areas. Biosphere 2 showed that it is possible to monitor very specific areas, very closely, at reasonable cost and, by extension, to monitor the entire planet. All the gases should be monitored, as we did at Biosphere 2.

It's especially important to discover and monitor all the processes taking place in Biosphere 1's soils. I had created a special soil in collaboration with the meticulous and outdoors-loving Dr. Bob Scarborough, a geologist. We collected, bagged and tagged soil samples to record their original state. The soil was monitored by Dr. Richard Harwood at Michigan State University, skilled in round-the-world agriculture, and lovingly cultivated by Sally Silverstone. Inspired by Darwin's classic paper on worms and their effect on the productivity of soils, I put five hundred thousand worms into that soil. We had to control insects and diseases by biological means to keep productivity high in the agriculture biome.

A proven sustainable soil, whose crops set records for production in a small area, without industrial chemicals, supported a beautiful landscape and eight human beings. That level of production would support about five thousand adults per square mile using efficient, chemical-free, machine-free (except for composter, small roto-tiller, and water pumps) agriculture, and can supply a complete diet. Only zinc and vitamin B-12 were slightly below requirements. Just recently (2004) did *Science* published an issue featuring soil and its cycles as a top priority for scientists.

In my contemplations during that spring of 1994, I dimly began to see taking a new step forward with Kathelin and the Theater of All Possibilities. "Frames" of vivid moments (based on conscious–altering memories of life-altering events) could replace the classic continuities and/or montage with their wildly meandering throughline of meanings climaxing in a catharsis of the past, the liberation of Space and the conquest of Time. Kathelin and I roamed further and further out in conversations. At last, we really played in our creative imagination on all the stages of the theater of all possibilities. We had thought ourselves free and at the tip of creativity with the Caravan of Dreams and Biosphere 2. Now we saw a thousand hang-ups we had not dared explore about The Dark Lady. Biosphere 2 had turned out to be a mystery epic. Atomic artist Tony Price e-mailed me on hearing of my leaving the site, saying, "This has been your greatest play." By the summer of 1994, a free man again, I visited Ornette Coleman several times, finally able to open myself to his direct teaching on sound. Ornette began by smashing my whole thought structure about "music" and finally got me to connect directly with the quality that he called harmolodics. To work his magic, Ornette made me sit cross-legged on the floor of his studio for four hours one night, plucking a violin until all my mental notions about "music" vanished. It was the same process by which Konrad Lorenz had smashed my mental structure about "biology" and put me in direct contact with the tempo and forms of life in action. The continuum of art and science, quality and quantity, dazzled me. Happiness returned – clear, sweet water soaking parched earth.

Re-entering into that resonating vibration of theater and music changed me again forever. I saw the secrets of the ethnosphere as clearly as I had seen those of the biosphere. I saw that the creation of the next biospheric model must integrate the ethnosphere. This was my great insight from Biosphere 2. The cybersphere and noösphere can only arrive (and the biosphere flourish) when the peoples of the ethnosphere unite themselves against being made into commodities and accomplices in the degradation of our home, the Earth. No culture owns Earth's atmosphere or its water, its diversity of genes and behavior. They all migrate and interact, and increasingly, so only a cybersphere, a universal feedback system on biospheric and technospheric events that any ethnic group and interested individuals and corporations can easily contact, will give people the means to combat the excesses of Global-Tech or create real freedom spaces.

The powers of art, music, drama, and poetry to transform experience into increased powers of perception, empathy, and apprehension of form re-entered me, a supreme gift. As an inadequate payback and to make my personal contribution to the cybersphere, I decided to write this memoir from a personal standpoint that does not claim any "objective truth." It would focus on what got my vision going, its ideas and its coming into being; tell about the hundreds of honest and brilliant individuals who created and fought for Biosphere 2. All who wish to, now and in the future, should be able to know something about the marvelous people around the planet who pitched in the effort with right good will and might and main.

The science and engineering of biospherics are essential tools to ecologists and ethnologists in the same way that cyclotrons are to physicists and space telescopes are to astronomers. They are of direct benefit to human understanding. We humans cannot progress much further without real knowledge about biospheric processes, and Biosphere 2 (and 3, 4, 5... to come) are major laboratories for obtaining that real knowledge.

Above all, I want present-day and future humans to know that Biosphere 2 existed, worked and was loved by me, my friends, colleagues and those millions of people who came into contact with it. Humans have the power to create new biospheres as well as to restore what has been ruthlessly plundered in Biosphere 1.

Biosphere 2 taught us much about how to build even better biospheres. As Goethe said, "Our friends show us what we can accomplish; our enemies show us what we must do." I will continue to watch and participate in the ongoing history of Biosphere 2 and its successors. Let this history enlighten whoever it may.

My Dream

TWENTY YEARS FROM NOW EVERY COUNTRY WILL have their own biosphere. Each biosphere will reflect the biomes in that country. For example, China would have a rainforest based on Yunnan, a coral reef from the China Sea, grasslands from the north, desert from Mongolia, a marsh from the coast, an agriculture from each of its various farmlands, and a human habitat representing the ultimate Shanghai or Beijing communication, technical, theater, cuisine, research, lay out. The same will be true for Great Britain, France, Russia, United States, Australia, Brazil, and all the others.

These biospheres will be open to public view; their scientific papers available to all. Those who manage the biosphere and those working in its supporting research centers, will form a school producing practical graduates who understand the biosphere. They constitute a cadre to work within governments, corporations, non-profit organizations, and popular movements to ensure that human progress makes a synergy with Earth's biosphere. Some of these graduates will become members of the core groups working to understand possibilities for biospheres on Mars, the Moon and in space. This work will contribute immensely to understanding the glorious uniqueness of our planet Earth, at least in this solar system.

These biospheres will demonstrate the ways that technosphere and ethnosphere work together with biosphere; from that comprehensive knowledge, a noösphere, a sphere of intelligence that unifies art, science, and humanity will emerge on planet Earth. These biospheres will be open to all; not special preserves for elites to create new bureaucracies. Farms, parks, cities, universities, and wilderness reserves will re-frame their ways to cooperate with and enhance our home, Earth's biosphere, with a new vision of a beneficial and profitable technosphere. They will act as cooperative agents of emergent evolution, and with a new understanding of ethnosphere, create a co-evolving system of cultures that upgrade their soils, flora, fauna, and humans' senses and imagination.

My intention was for Biosphere 2 to be a gift to humanity. Everyone who came to Biosphere 2 saw exactly what I saw; we hid nothing. We took no money from the United States or any state government. We stayed independent of any professional organization. We pushed no ideology to gain support of their followers. My scientific and artistic friends freely contributed their time and advice; some contributed money and graduate students to help secure vital data. All of us in the top management took just enough pay to cover our costs of living as simply and biospherically as we knew how. We have written and published all the relevant papers on how a biosphere operates to transmit what we learned. This book completes my task on that line.

I called Biosphere 2 in one of my books, the "Human Experiment." There were many questions about what humans would do with it. What happened at Biosphere 2 after I left in '94 was: the facility no longer has people living inside it. It's been instructive to see the evolution of the project as it was designed to last for one hundred years. The first group of university scientists replaced the sustainable agriculture system with cottonwood trees to study growth rates under different amounts of CO_2. The facility was re-engineered so it is no longer totally materially-closed – and most of the research has focused on individual biomes, not the total system. Good research, but not taking advantage of the unique potentials a mini-biosphere offers. But the story is far from over – the University of Arizona has recently taken over management of the project in a present world where there is growing concern about depletion of natural resources and the biosphere's ability to sustain itself under the increased impacts of the technosphere.

"The Human Experiment" confirms that our present thinking and behaving about biospheres tends to be destructive of the life system.

There is not just a CO_2 problem on the planet. There is a biospheric problem. An increasing CO_2 content in the air is but a symptom, not the disease. The real problem is the entrenched attitude that the biosphere is an "environment" to exploit rather than enrich. UNESCO has a program called "Man and the Biosphere." Inaccurate. We humans are part of the biosphere; we are not separate. Our own actions cause the biosphere problem, and our actions stem from defects in our ideas, institutions, and values. Where is the Secretary of the Biosphere in the United States government?

As to values, artists, scientists, philosophers (except in Russia), and political leaders ignore biospherics, while some gain fortune and fame attacking CO_2's increase in the atmosphere, which of course is inevitable with increasing number of humans generating increasing numbers of CO_2 producing equipment. Concentrating effort on CO_2 is like curing a headache in someone dying of cancer (an uncontrolled growth heedless of the total system it is growing on); it relieves a little pain but does not prevent the death of the patient. Governments, corporations, and the vast majority of citizens ignore the biosphere. Species are driven to extinction as if they didn't play vital roles in the biospheres's equilibrium.

In my original design for Biosphere 2, part of the property was to become an international biospheric university. This university we envisioned would be a consortium of Oxford, Harvard, Tokyo, Moscow universities, and other institutions. The first country to do this will lead the world in a new scientific era. In my dream, my own country wakes up and accomplishes this magnificent feat for all people.

My full dream: Earth's biosphere will flourish with humanity as its star performer; the technosphere will support the biosphere as its fragile, unique base from which to intellectually and physically explore the universe; the ethnosphere's cultures will unite to stop destruction of magnificent age-old biospheric systems for evanescent exploiters to squander on conspicuously idiotic consumption.

A Sweet Blue Grape Under a Yellow Sun in a Green Garden

MY COLLEAGUES AND I CONTINUE WORK ON BIOSPHERIC THEORY, publishing papers, attending and giving conferences in Les Marronniers, in London, and at Synergia Ranch in New Mexico. We give talks in Rome, Tokyo, Moscow, Beijing, London.

Biospheric Design, a division of Global Ecotechnics, is now run by Mark Nelson, Bill Dempster, and me. It consults with Margaret Augustine on architecture and other matters. Global Ecotechnics, started in July 1994, free and clear, is engaged in six biospheric projects. Margaret Augustine has her architecture office at Synergia Ranch. Dr. Mark Nelson continues to skillfully chair the annual Institute of Ecotechnics conference, which Deborah Parrish Snyder organizes. IE conferences continue to attract scientists and thinkers from all over. Mark travels the world as a top expert on wastewater recycling, and sparkplugs our Australian savannah project. Marie Harding has created a new atelier, inspired by Cézanne's, for her painting. She operates part of Synergia Ranch as a center of innovation for lively groups to hold their conferences.

As for me, like all of us humans I came from and will return to the biosphere, which itself will co-evolve billions of better adapted *me's*. Living every day as part of Earth's unique biosphere creates many a brimming moment, like when a young boy could be found eating a sweet blue Concord grape on a hot August afternoon on the Oklahoman prairie. His grandfather's grape vine sprawled along the garden fence, fertilized by human wastes. The boy thoughtfully chewed the grape-skin after tracing a finger trail through its bloom, tasting the flesh and juice. Then, crushing the seeds with his molars, he took a long satisfying swallow and registered the drop of goodness in his stomach. He licked his lips, selected the next sphere, making the dexterous twist-off with thumb and fingers, and would not stop until the cluster of blue spheres under the yellow sun, in a garden of green, dropped its skeleton to the ever-richer soil from which it emerged.

Acknowledgments

GREAT THANKS ARE DUE TO MANY, but especially to my grandfather Brune Wall, my grandmother Kate Wall, to my great aunt, Jenny Wall, and my father and mother who taught me how to live in and love Earth's biosphere.

To Margaret Augustine, Marie Harding, William Dempster, Kathelin Hoffman Gray, Edward Bass, Mark Nelson, and Robert Hahn who helped me set Biosphere 2 into motion.

To Bucky Fuller, Eddie McKee, Jack MacCauley, James Head, Enrico Fermi, Konrad Lorenz, Bernd Lötsch, Richard Evans Schultes, Eugene Odum, Howard Odum, Keith Runcorn, Ghillean Prance, Oleg Gazenko, Evgenii Shepelev, Josef Gitelson, Roy Walford, Albert Hibbs, Rusty Schweickart, Niles Eldredge, Walter Orr Roberts, Richard Dawkins, Lynn Margulis, Ralph Abraham, Bill Chaloner, John Marsden, Joseph Allen, all whose love of life and science inspired my thought about Biospheres.

To William Burroughs, Brion Gysin, Ornette Coleman, Gerald Wilde, Lawrence Durrell, Tambimuttu, Joe Cino, Bruce Goff, Hassan Fathy, Paolo Soleri, who aroused my artistic conscience.

To those powerful sages, shamans, and unknown teachers from Titicaca to Sierra Madre del Oeste to Sangre de Cristo to Kyoto to Himalayas to Tiger Tops to Delhi to Herat to Istanbul to Ile-Ife to Tangiers to Haight-Ashbury to The Village to the Rive Gauche to Bloomsbury to Moscow to Cairo to Isfahan to the Upper Nile, who, desiring no recognition, labored to awaken my sleeping Mind before it was too late see what I have been lucky enough to see.

To Deborah Parrish Snyder without whom this book would never have been written, revised, or published.

Note from the Publisher

A book should be designed like a crystal-clear goblet ... calculated to reveal the beautiful thing it was meant to contain.
BEATRICE WARDE

THERE ARE MANY PEOPLE WHO HELPED make this book possible. Foremost, Aryln Eve Nathan, the book's designer, who is responsible for the clarity of this goblet. During the process of production, I learned from her many things about the fine art of book design. And Linda Sperling, who edited the manuscript with grace and skill, working closely with the author to bring his voice and story to these pages, and who helped with this production every step of the way. Thank you.

And thanks to Corinna MacNeice who spent hours in the archive with me going through hundreds of images, some in dusty old boxes, to help find and scan the photos in this book. Also to the publishing editorial and production team: Monique Sofo, Ari Phillips, Joan Schweighardt, Ellen Kleiner, Dina McQueen. Special thanks to early editorial work on this project from Parvati Markus; to Colin Brown and Hugh Elliot for their very helpful inputs; to Peter Ellzey for technical support, and to John Corey for keeping the computers from crashing.

And to the other friends and colleagues who jumped in when needed to review, fact check, spell check, write promotional copy, look for photographs and make coffee. Among them, Mark Nelson, Kathelin Gray, Chili Hawes, Gessie Houghton, and Maria Golia.

I'd also like to thank our printer, Arizona Lithographers in Tucson, and its owner, John Davis, for helping to support this publication.

DEBORAH PARRISH SNYDER

Photo Credits

IT HAS BEEN AN AMAZING JOURNEY going through John's archive to find so many stunning images with which to illustrate his extraordinary life. We thank all the photographers who captured these moments over the years for their permission to reproduce these images.

Four photographers contributed most of photographs that appear throughout this book, too numerous to identify each by page: Gill C. Kenny magnificently captured the construction, launching and architecture of Biosphere 2 from 1989–1993; Marie Harding and Robert "Rio" Hahn documented all the projects and events described in this book over the course of forty years; and I, myself, Deborah Parrish Snyder, began to capture this history through the lens from 1984 onward.

Other photographers include: Rafael Aisner: 42,54 (color section) 3; Abigail Alling: 212, 222 (color section) 4, 16; Gonzalo Arcila: 29, 127, 168, (color section) 2, 110; Brian Blauser: 106; Ira Cohen: 105, 107; Kathelin Hoffman Gray: 85; Jonathan Greet: 83, (color section) 5; Tom Lamb: 163; Michel Lippitsch: 300; Scott McMullen: (color section) 14; Bernd Lötsch: 285; Chuck McDougal: 62; Peter Menzel: 154, (color section) 9, 10, 15; C. Allan Morgan: 124, 174, 204, 272 (color section) 11; Elena Monforte: 144; Roger Ressmeyer: (color section) 11; Andrés Rúa: 92; Mike Stokolos: 224, 246; Jeff Topping: 188; Thrity Vakil: dedication page, 92, (color section) 7. We were unable to identify photographers of some images. Our apologies if your photo credit was missed.

Photographs of Biosphere 2 from John Allen's archive, reprinted with permission. Thanks to Roy Allen for providing early family photos. Photos of Theater of All Possibilities, courtesy of TAP. Jazz and Blues Mural at Caravan of Dreams © Zara Kriegstein. Images of Bio3 in Russia, courtesy of Institute of Biophysics.

Footnotes

1   Allen, J.P. 2003 "Ethnospherics: Origin of Human Cultures, Their Subjugation by the Technosphere, the Beginning of an Ethnosphere, and Steps Needed to Complete the Ethnos phere." *Ethics in Science and Environmental Politics* (ESEP) (http://www.int-res.com/articles/esep/2003/E29.pdf)

2   James Turpin and Al Hirshberg, *Vietnam Doctor: The Story of Project Concern*, McGraw Hill, NY, 1966.

3   John Allen and Mark Nelson, *Space Biospheres*, Synergetic Press, Oracle, AZ, 1986, 1989.

4   NASA was (and still is) sold on hydroponics and its simplified, yet expensive systems for semi-closed systems; but our work and others show hydroponic systems have severe build-up problems in a closed life system.

5   IE was started in the U.S. in 1969, incorporated in 1973, became a UK registered institute in 1985; became a UK charity in 2000 and a U.S. non-profit corporation in 2007.

Links

Institute of Ecotechnics: www.ecotechnics.edu.
Research Vessel *Heraclitus*: www.rvheraclitus.org
Synergia Ranch: www.synergiaranch.com
Biosphere 2: www.biospherics.org
Global Ecotechnics Corporation: www.globalecotechnics.com
Wastewater Gardens International: www.wastewatergardens.com

John Allen publications, speeches, projects around the world at: www.meandthebiospheres.com.

Bibliography

NON-FICTION

Allen, John, *Biosphere 2: The Human Experiment*, Viking/Penguin Books, 1991.

Allen, John. *Succeed: Structuring Managerial Thought*, Synergetic Press, 1988.

Allen, John and Mark Nelson, *Space Biospheres*, Synergetic Press, 1986; Orbit Publications, hardcover. 1987. Russian translation by Progress Publishers, Moscow, 1991.

NOVELS AND SHORT STORIES
(under pen name Johnny Dolphin):

Trilogy of the Sixties:
    Liberated Space, Synergetic Press, 2000.
    Journey Around An Extraordinary Planet, Synergetic Press, 1990.
    Thirty-nine Blows on a Gone Trumpet, Synergetic Press, 1987.
My Many Kisses, Synergetic Press, 1998. (Short Stories)

POETRY (by Johnny Dolphin)

    Off the Road, Synergetic Press, 1998.
    Wild, Synergetic Press, 1992. (Collected Poems and Short Stories)
    The Dream and Drink of Freedom, Synergetic Press, 1988.

ADAPTATIONS OF PLAYS

Faust; Gilgamesh; Oedipus at Colonus; Life is a Dream; Milarepa; and *Deconstruction of the Countdown*, a play adapted with Kathelin Hoffman Gray from William S. Burrough's writings. (*Deconstruction of the Countdown: A Space Age Mythology* was published in abridged form in *Poetry London Apple Magazine*, No. 2, Editor: Tambimuttu, Editions Poetry London, 1982.)

John Dolphin Allen and Kathelin Hoffman Gray outside the Caravan of Dreams in Fort Worth, Texas.

ORIGINAL PLAYS

Tamarand; McNeckel's Folly; Marouf the Cobbler; The Tin Can Man; The Guru; Metal Woman; Billy the Kid; and the *Carneval of the Seven Sins, Cyberspace 1; Cyberspace 11* with Kathelin Hoffman Gray. (Some of these were published in *The Collected Works of the Caravan of Dreams Theater Volumes 1 and 11*, Synergetic Press, 1983/84.)

FILMS

1987   Creative consultant, prize-winning film, *Omette: Made in America*, based on the life of jazz musician Ornette Coleman, Caravan of Dreams Productions, Fort Worth.

1987   Documentary film producer and narrator, *Journeys to Other Worlds*, a series of eleven cultural documentary films, Caravan of Dreams Productions in cooperation with Zagreb Television, Yugoslavia

1986   Documentary film producer, prize-winning film, *Dark Planet*, Caravan of Dreams Productions in cooperation with Zagreb Television, Yugoslavia.

SCIENTIFIC PAPERS

Allen, J.P., M. Nelson and A. Alling, 2003. "The legacy of Biosphere 2 for the study of biospherics and closed ecological systems," paper presented at the World Space Congress, COSPAR general assembly, Houston, Texas, October 2002, *Advances in Space Research* 31(7): 1629-1640.

Allen, J.P. 2000. "Artificial Biospheres as a Model for Global Ecology on Planet Earth," *Life Support and Biosphere Science*, Vol. 7, No. 3.

Allen, J.P. and M. Nelson, 1999. "Biospherics and Biosphere 2, Mission One (1991-1993)," *Ecological Engineering* 13 (1-4): 15-26.

Zelko Malnar, co-director and cameraman of the "Journey's to Other Worlds" film series, with John on location in Rajasthan, India.

Allen, J.P. 1997. "Biospheric Theory and Report on Overall Biosphere 2 Design and Performance During Misson One (1991-1993) *Life Support and Biosphere Science*, Vol. 4, No. 3/4.

Allen, J.P., 1991. "An Historical Overview of the Biosphere 2 Project." Workshop on Biological Life Support Technologies: Commercial Opportunities, Mark Nelson and Gerald Soffen (ed.), NASA Conference Publication, NASA Office of Management, Scientific and Technical Information Division, Washington D.C., pp. 12-22, 1990. Also published in *Biological Life Support Systems: Proceedings of the Workshop on Biological Life Support Technologies: Commercial Opportunities*, Mark Nelson and Gerald Soffen (ed.), Synergetic Press.

Allen, J.P. 1989. "Ecology and Space." Paper published in the *Japanese Society for Biological Sciences in Space* publication series, Vol. 3, No. 1, Tokyo, 1989, and reprinted in *Lunar Habitation* Vol. 1, by Japanese Macro-Engineer Society and Lunar Habitation Institute, Tokyo.

Alling, A., M. Van Thillo, W.F. Dempster, M. Nelson, S. Silverstone and J.P. Allen, 2005. "The Mars On Earth® Project: Lessons Learned from Biosphere 2 and Laboratory Biosphere Closed Systems Experiments," *Biological Sciences in Space*, 19(4): 250-260.

Dempster, W.F., M. Nelson and J.P. Allen, "Atmospheric dynamics of combined crops of wheat, cowpea, pinto beans in the Laboratory Biosphere closed ecological system." Paper presented at COSPAR 2006 in Beijing, in press *Advances in Space Research*.

Dempster, W.F., Allen, J.P. Alling, A., Nelson, M., Silverstone, S. and M. Van Thillo, 2005. "Atmospheric dynamics in the 'Laboratory Biosphere' with wheat and sweet potato crops," *Advances in Space Research* 35(9):1552-1556.

Dempster, W.F., A. Alling, M. Van Thillo, J.P. Allen, S. Silverstone and M. Nelson, 2004. "Technical review of the Laboratory Biosphere closed ecological system facility," presented at COSPAR/IAF meeting, Houston, October 2002, *Adv. Space Research*, 34: 1477-1482.

Morowitz, H., J.P. Allen, M. Nelson and A. Alling, 2005. "Closure as a Scientific Concept and its Application to Ecosystem Ecology and the Science of the Biosphere," paper presented at COSPAR, Paris, 2004; *Advances in Space Research* 36(7): 1305-1311.

Nelson, M.; Allen, J.P.; and Dempster, W.F., 2008. "Modular Biospheres – a New Platform for Education and Research," *Advances in Space Research* 41(5): 787-797.

Nelson, M.; J.P. Allen and W.F. Dempster, 2008. "Integration of lessons from recent research for 'Earth to Mars' life support systems," *Advances in Space Research* 41(5): 675-683.

Nelson, M., W.F. Dempster, J.P. Allen, S. Silverstone, A. Alling, 2008. "Cowpeas and pinto beans: Performance and yields of candidate space crops in the Laboratory Biosphere closed ecological system," *Advances in Space Research* 41(5): 748-753.

Nelson, M., W.F. Dempster, S. Silverstone, A. Alling, J.P. Allen and M. Van Thillo, 2005. Crop Yield and Light/Energy Efficiency in a Closed Ecological System: Laboratory Biosphere experiments with wheat and sweet potato, *Advances in Space Research* 35(9): 1539-1543.

Nelson, M., A. Alling, W.F. Dempster, M. Van Thillo, and J.P. Allen, 2002. "Integration of wetland wastewater treatment with space life support systems." Paper presented at COSPAR conference, Warsaw, Poland, July, 2000, *Life Support and Biosphere Science* 8(3/4): 149-154.

Nelson, M., J.P. Allen, A. Alling, W.F. Dempster and S. Silverstone, 2003. "Earth applications of closed ecological systems: relevance to the development of sustainability in our global biosphere." Paper presented at the World Space Congress, COSPAR general assembly,

John speaking at the Royal Society meeting in 1987, London.

Houston, Texas, October 2002, *Advances in Space Research*, 31(7): 1649-1656.

Nelson, M., T.L. Burgess, A. Alling, N. Alvarez-Romo, W.F. Dempster, R.L. Walford and J.P. Allen,1993. "Initial Results from Biosphere 2: A Closed Ecological System Laboratory," *BioScience* 43(4): 225-236.

Salisbury, F.B., W.F. Dempster, J.P. Allen, A. Alling, D. Bubenheim, M. Nelson and S. Silverstone, 2002. "Light, Plants and Power for Life Support on Mars." Paper presented at NASA Mars Ecosynthesis workshop, Santa Fe, New Mexico, September 2000, *Life Support and Biosphere Science*, 8(3/4):161-172.

Silverstone, S., M. Nelson, A. Alling, J.P. Allen, 2003. Development and research program for a soil-based bioregenerative agriculture system to feed a four person crew at a Mars base. *Advances in Space Research*, 31(1): 69-75.

ETHNOSPHERIC PAPERS

Allen, John, 2003. "Ethnospherics: Origins of human cultures, their subjugation by the technosphere, the beginning of an ethnosphere, and steps needed to complete the ethnosphere." *Ethics in Science and Environmental Politics* (ESEP):7-24.

Allen, John, 2000. "The Evolution of Humanity: Past, Present and Possible Future – Review of humanity's taxonomic classification and proposal to classify humanity as a sixth kingdom, Symbolia." *Duversity Newsletter*, No.4 (edited and published by the British thinker, A.G.E. Blake).

For other books about biospherics and Biosphere 2 published by Synergetic Press, visit www.synergeticpress.com.